Art in Science Museums

Art in Science Museums brings together perspectives from different practitioners to reflect on the status and meaning of art programmes in science centres and museums around the world.

Presenting a balanced mix of theoretical perspectives, practitioners' reflections and case studies, this volume gives voice to a wide range of professionals, from traditional science centres and museums and from institutions born with the very aim of merging art and science practices. Considering the role of art in the field of science engagement, the book questions whether the arts might help curators to convey complex messages, foster a more open and personal approach to scientific issues, become tools of inclusion and allow for the production of totally new cultural products. The book also includes a rich collection of projects from all over the world, synthetically presenting cases that reveal very different approaches to the inclusion of art in science programmes.

Art in Science Museums should be of great interest to academics, researchers and postgraduate students working in the fields of museum studies, cultural heritage management, material culture, science communication and contemporary art. It should also be essential reading for museum professionals looking to promote more reflective social science engagement in their institutions.

Camilla Rossi-Linnemann is currently International Partnership Coordinator at the National Museum of Science and Technology Leonardo da Vinci in Milan, where she focuses on international projects and collaborations for visitor engagement and cultural diplomacy, especially based on the interrelation between science, technology and art.

Giulia de Martini is currently Head of Research at TheFabLab, a digital manufacturing laboratory, where she studies the impact of innovation in society and spreads awareness about the technological and human competences involved in the fourth industrial revolution, focusing on the connections between different fields of knowledge.

Routledge Research in Museum Studies

The Personalization of the Museum Visit
Art Museums, Discourse, and Visitors
Seph Rodney

Narratives of Vulnerability in Museums
American Interpretations of the Great Depression
Meighen Katz

Museum and Gallery Publishing
From Theory to Case Study
Sally Hughes

Museums and Centers of Contemporary Art in Central Europe after 1989
Katarzyna Jagodzińska

Exhibitions as Research
Experimental Methods in Museums
Edited by Peter Bjerregaard

Museum Objects, Health and Healing
The Relationship between Exhibitions and Wellness
Brenda Cowan, Ross Laird and Jason McKeown

Museums of Language and the Display of Intangible Cultural Heritage
Edited by Margaret J-M Sönmez, Maia Wellington Gahtan and Nadia Cannata

Art in Science Museums
Towards a Post-Disciplinary Approach
Edited by Camilla Rossi-Linnemann and Giulia de Martini

For more information about this series, please visit: https://www.routledge.com/Routledge-Research-in-Museum-Studies/book-series/RRIMS

Art in Science Museums

Towards a Post-Disciplinary Approach

Edited by Camilla Rossi-Linnemann
and Giulia de Martini

Routledge
Taylor & Francis Group
LONDON AND NEW YORK

First published 2020
by Routledge
2 Park Square, Milton Park, Abingdon, Oxon OX14 4RN

and by Routledge
52 Vanderbilt Avenue, New York, NY 10017

Routledge is an imprint of the Taylor & Francis Group, an informa business

First issued in paperback 2021

British Library Cataloguing-in-Publication Data
A catalogue record for this book is available from the British Library

Library of Congress Cataloging-in-Publication Data
A catalog record for this book has been requested

ISBN: 978-1-138-58952-0 (hbk)
ISBN: 978-1-03-208297-4 (pbk)
ISBN: 978-0-429-49159-7 (ebk)

Typeset in Sabon
by Apex CoVantage, LLC

Contents

Figures

Contributors

Editors

Camilla Rossi-Linnemann is currently International Partnership Coordinator at the National Museum of Science and Technology Leonardo da Vinci in Milan, where she focuses on international projects and collaborations for visitor engagement and cultural diplomacy, especially based on the interrelation between science, technology and art. Her expertise and interests revolve around the development of strategies for visitor engagement and cultural diplomacy. She has worked on numerous EU projects and collaborative education projects, and published with respect to the inclusion of specific audiences such as visually impaired or very young visitors. Her early involvement in physics education has fuelled a passion for science and for thinking about ways to develop real participatory citizenship. She holds an MA in Museum Studies (Leicester University – UK) and a BA in Art History (University of Milan – Italy).

Giulia de Martini is Head of Research at TheFabLab (Milan, Italy), a digital manufacturing laboratory, where she studies the impact of innovation in society and spreads awareness about the technological and human competences involved in the fourth industrial revolution, focusing on the connections between different fields of knowledge. In her former position at the National Museum of Science and Technology in Milan, she performed her research in the area of Visitors Studies and Evaluation. As a member of the education staff, she developed programmes merging the opportunities offered by collections and interactive labs to engage visitors. A physicist who graduated in Milan, Italy, she holds a Master in Science Communication at SISSA (Trieste, Italy) where she first approached museum studies.

In 2014 Camilla Rossi-Linnemann and Giulia de Martini co-founded SPARK Science&Art, leading projects that bring together scientists and artists in the double attempt to reach wider communities and to bring innovation in both the fields of humanities and sciences.

Preface author

Roger Malina is an astronomer and editor. He is a Distinguished Professor of Art and Technology at the University of Texas, Dallas, where he is developing Art-Science R&D and Experimental publishing research. Former Director of the Observatoire Astronomique de Marseille Provence. His specialty is in space instrumentation; he was the Principal Investigator for the NASA Extreme Ultraviolet Explorer Satellite at the University of California, Berkeley. He also has been involved for 25 years with the Leonardo organisation, whose mission is to promote and make visible work that explores the interaction of the arts and sciences and the arts and new technologies. Since 1982 he has been the Executive Editor of the Leonardo Publications at MIT Press. More recently he has helped set up the Mediterranean Institute for Advanced Studies (IMERA) and is co-chair of the ASIL (Arts, Sciences, Instrumentation and Language) Initiative of IMERA, which hosts artists in residence in scientific research laboratories of the Marseille region. He is an elected member of the International Academy of Astronautics.

Contributors

Bergit Arends is an academic researcher and curator of contemporary art. She creates interdisciplinary research environments, develops research capacity within museums, and explores the interferences of the arts and the sciences, focusing on environmental humanities and visual art.

She is currently Deputy Head of Collection Care Research at Tate, London. Previously she was Acting Head of Research at Science Museum Group. Her PhD thesis of 2017, completed at Royal Holloway, University of London, is titled 'Contemporary art, archives and environmental change in the age of the Anthropocene'. Bergit curated the Kunst/Natur program at the Natural History Museum, Berlin (2016–2018) and was Curator of Contemporary Art at the Natural History Museum, London (2005–2013).

The programme at NHM included international artists' residencies, commissions, and exhibitions. Among these exhibitions were the touring group shows Galápagos (2012–2013) and After Darwin: Contemporary Expressions (2009); as well as Lucy + Jorge Orta: Amazonia; Mark Dion: Systema Metropolis (2007); and The Ship: The Art of Climate Change (2006). For the museum's Treasures Gallery, she commissioned the permanent installation TREE (2009) by Tania Kovats to mark the 150 years of the publication of "On the Origins of Species" and the 200th anniversary of Darwin's birth. In August 2016 she participated in the first workshop in Karongi, Rwanda, on cultural heritage protection and environmental conservation in the African context as part of the international Culture in Crisis programme by the V&A and the Yale Institute for the Preservation of Cultural Heritage, in collaboration with the National Museums of

Rwanda and the Museum für Naturkunde, Berlin. She studied Curating at the Royal College of Art, London.

Ken Arnold has been Director of Medical Museion and Professor at Copenhagen University since June 2016. He also continues as Creative Director at Wellcome, where he oversees international cultural projects. He arrived at Wellcome in 1992, where he spearheaded a number of multidisciplinary cultural initiatives, culminating in 2007 with the establishment of Wellcome Collection, in which he directed the programmes for a decade. He regularly writes and lectures on museums and on contemporary intersections between the arts and sciences. His book "Cabinets for the Curious" (Ashgate, 2006) explored what can be learned from looking back at England's earliest museums. He is currently researching a book about the increasingly important cultural role for these enduring institutions.

Andrea Bandelli is a scholar in science communication and the executive director of Science Gallery International. He has developed, managed, evaluated and directed several international projects on science, art, democracy and public participation in collaboration with Ecsite, the European network of science centers and museums and with science museums, government organisations and universities across Europe, USA, South Africa and Brazil.

He has a PhD in Science Communication from the Vrije Universiteit in Amsterdam on scientific citizenship and the role of public participation in science centers, and is a Cultural Leader at the World Economic Forum. He is a member of the board of the Deutsches Museum in Munich.

Adam Bencard is Associate Professor in medical humanities at the Medical Museion, University of Copenhagen, and also affiliated with The Novo Nordisk Foundation Center for Basic Metabolic Research. His research interest includes the philosophical and cultural ramifications of contemporary microbiome research and what it means to be human in a postgenomic world. He also researches science communication in museum contexts and museology, as his position is split evenly between research and curatorial practice. He has curated several exhibitions at the Medical Museion, including the award-winning *Mind the Gut* (alongside Associate Professor Louise Whiteley) which opened in the fall of 2017. He is currently working on a research project at the intersection of art and science with regards to microbiome research, titled "You are not alone – microbiomes as model ecologies in art and science".

Ian Brunswick is the Futures Curator at the Smithsonian's Arts & Industries Building, and was previously Head of Programming at Science Gallery, Trinity College Dublin. Since 2010, he led the development of Science Gallery exhibitions including *Illusion, Design and Violence, In Case of Emergency* and *Secret*. He has overseen projects with partners including

University of Cambridge, the Museum of Modern Art in New York, the Irish Health Research Board, Science Foundation Ireland and the Wellcome Trust.

He sits on the board of the Irish Museums Association, and was previously a member of the ECSITE annual conference programme committee and the jury for the Bio-Art and Design Award. He holds a BA in Comparative Literature from the University of Wisconsin-Madison and an MSc in Science Communication from Dublin City University.

Oron Catts is the Director of SymbioticA, The Centre of Excellence in Biological Arts, School of Human Sciences, The University of Western Australia. He is an artist, researcher, designer and curator whose pioneering work with the Tissue Culture and Art Project, which he established in 1996, is considered a leading biological art project. In 2000 he co-founded SymbioticA, which later was awarded the inaugural Prix Ars Electronica Golden Nica in Hybrid Art (2007), the WA Premier Science Award (2008), and became a Centre for Excellence in 2008. In 2009 Catts was recognised by Thames & Hudson's "60 Innovators Shaping our Creative Future" book in the category "Beyond Design", and by Icon Magazine (UK) as one of the top 20 Designers, "making the future and transforming the way we work". Catts was a Research Fellow at Harvard Medical School, a visiting Scholar at the Department of Art and Art History, Stanford University, and a Visiting Professor of Design Interaction (2009–2012), and a Professor at Large in Contestable Design (2015–2017) at the Royal College of Arts, London. In 2012–2013 he set up a biological art lab called Biofilia – Base for Biological Art and Design, at the School of Art, Design and Architecture, Aalto University, Helsinki, where he was a Visiting Professor. Catts' ideas and projects reach beyond the confines of art; his work is often cited as inspiration to diverse areas such as new materials, textiles, design, architecture, ethics, fiction, and food. His work was exhibited and collected by museums such as MoMA NY, Mori Art Museum, NGV, GoMA, Yerba Buena Center for the Arts, San Francisco, Ars Electronica, National Art Museum of China and more.

John Durant has spent his entire career working at the interface between science and technology and the wider community. He was Assistant Director of the Science Museum, London, between 1989 and 2000, where he oversaw the development of the Wellcome Wing. From 2000–2005 he was Chief Executive of At-Bristol, a new science and natural history museum in the west of England. In 2005, he was appointed as Director of the MIT Museum and Adjunct Professor in the Science, Technology & Society Program at MIT. Here, he leads the development of the MIT Museum as a place where the wider community can engage with research and innovation, past and present. In 2007, he led the creation of the Cambridge Science Festival, the first citywide festival of science and technology in the U.S. In 2008, he was a founder Fellow of the Noyce Foundation Science Center Leadership Initiative.

Nuno Ferrand de Almeida is Full Professor of Evolutionary Biology at the Faculty of Sciences, University of Porto, Scientific Coordinator of CIBIO, a research centre in biodiversity and genetic resources based in Portugal with more than 300 researchers and Director of InBIO – Research Network in Biodiversity and Genetic Resources. He is also invited professor at the University of Johannesburg in South Africa, where he teaches genetics and evolution. His research interests comprise evolutionary biology, particularly in genetic diversity patterns of natural populations, ecology, conservation, domestication and speciation. Since 2011, as general director and in partnership with the Portuguese National Agency for Scientific and Technological Culture – Ciência Viva, he started an unparalleled project of conceiving and implementing a new Museum of Natural History and Science in the city of Porto, bringing together, for the first time ever in the country, a university museum and a science centre, where art plays a fundamental role.

Maria João Guimarães Fonseca is Interim Executive Coordinator and Head of Communication at the Natural History and Science Museum of the University of Porto. With degrees in Applied Animal Biology and Biology Teaching and a PhD in Science Education and Communication, she has extensive experience in the design, implementation and evaluation of educational activities in formal and non-formal learning environments as well as in institutional communication. Among her interests are topics such as science engagement, the convergence of art, science and technology in the promotion of scientific culture, science education in informal environments, especially museums and science centers, active learning methodologies, and assessment and evaluation of educational and communication activities.

Pietro Greco is a scientific journalist and writer. Founding partner of the science centre Città della Scienza in Naples, he collaborates with both RAI (Italian National Television) and numerous newspapers and with various universities in the training of young science communicators. He is author of over 25 monographs. On the subject of art and science, he wrote *L'astro narrante. La Luna nella scienza e nella letteratura italiana* (Springer Italia, 2009); *L'universo a dondolo. La scienza nell'opera di Gianni Rodari* (Springer Italia, 2010); and *Galileo, l'artista toscano* (Springer Italia, 2013; English version also with Springer). He was also the curator of *ArmonicaMente. Arte e scienza a confronto* (Mimesis, 2014). With Bruno Arpaia he published *La cultura si mangia!* (Guanda, 2013), devoted to cultural politics and therefore, in Italy, to art and science.

Samar D. Kirresh is a Principal Researcher at the Science Studio – A.M. Qattan Foundation – Ramallah-Palestine. She holds a BSc in Biology and

Medical Technology and an MA in Science Education. She is self-taught in arts and crafts. She had an active career in teaching for 12 years at College Des Frere Jerusalem, Head of Science Department for 6 years. She joined Waleed and Helen Qattan Science Education Project at AMQF in 2012 and was on the team that established the Science Studio. She is member of the Advisory Board for the College of Education at Birzeit University since 2011, the International Advisory Board for the Hands-on Science International Conference since 2014 and has been part of the team leading Maker Space Ecsite since 2016. Samar's work and research focuses on developing empowerment and public programmes. She has an extended experience in designing interactive integrated learning experiences and environments. She has been leading *Science Days in Palestine*, the *Science Film Festival* and *Nowar Nisan* festivals in cooperation with local and international organisations since 2013. She has an experience in exhibit development and fabrication with contributions in two exhibitions; *Light and Shadow* 2017, and *Motion and Stillness* 2018. She established the *Tinkering Room* as an integral space within the Science Studio. She founded *Washem*, an art project on motifs inspired from exploration of the Palestinian landscape and culture. She also co-founded *Art Market Jerusalem* a cultural platform for Palestinian artisans and designers to share products and expertise with the community.

Christopher Lindinger has been dealing with the social impact of new technologies and the cultures of innovation associated with them for more than two decades. He works as Director of Research & Innovation at Ars Electronica Futurelab, Linz Austria. Christopher Lindinger studied computer science and mathematics at Johannes Kepler University in Linz and worked on supercomputer visualisation of abstract mathematical models as a visiting researcher at the University of Illinois at Chicago. In 1997, his interest in interdisciplinary research and digital culture brought him to Ars Electronica, where he was a key protagonist in establishing the Ars Electronica Futurelab in the late 1990s. As co-director of the Ars Electronica Futurelab, he has managed the media lab's research activities since 2002. His work currently focuses on collaborative R&D with the private sector and scientific institutions. Together with international partners, he develops digital creativity support tools, and investigates the manifold potential of disruptive technologies such as VR, robotics and blockchain in social and corporate transformation processes. In addition to his work at the Ars Electronica Futurelab, he is a consultant to cities and government agencies worldwide on how to nurture creative industries.

Marina McDougall is a curator based in Philadelphia who specialises in interdisciplinary artistic approaches in public learning environments. She was most recently the Executive Director of the Philadelphia Art Alliance at the University of the Arts, one of the first multidisciplinary arts forums in the U.S. Previous to this role, McDougall was the Founding Director of the Exploratorium's Center for Art & Inquiry, where she developed the arts strategy for the renown public learning laboratory and oversaw the Exploratorium's Artist-in-Residence and Over the Water programs. McDougall was a co-founder of the Studio for Urban Projects; and served as the Associate Director for Program; and Curator of Art and Design at the California College of Arts and Crafts Institute for Art (today known as the Wattis Institute for Contemporary Art). She has taught as an Adjunct Professor in the California College of the Arts Curatorial Practice and Graduate Fine Arts Programs. McDougall is a trustee of the Museum of Jurassic Technology (Los Angeles), and serves on the board of Slought (Philadelphia), and the Seed Fund (San Francisco). She is a co-editor of *The Marvelous Museum: Orphans, Curiosities & Treasures: A Mark Dion Project* (Chronicle Books, 2006); and *Science is Fiction: The Films of Jean Painleve* (MIT/Brico Press, 1999). McDougall's long-term project *The Garden of Forking Paths* involves drawing upon the history of gardens to shape new forms of garden experience.

Luís Mendonça holds a degree and a PhD in design from the University of Porto. He is a Professor at the Faculty of Fine Arts, University of Porto, where he teaches design. Along the years, Luís has been engaged in the development of inspiring illustration and museography projects, being in charge of numerous exhibitions, many of which have been designed for the Natural History and Science Museum of the University of Porto and, in particular, the Hall of Biodiversity, whose aesthetical solutions were coined under his coordination.

Rui Mendonça holds a degree and a PhD in design from the University of Porto. He is a Professor at the Faculty of Fine Arts, University of Porto, where he teaches design. Through the years, Rui has been engaged in the development of innovative editorial and museography projects, being in charge of numerous exhibitions, many of which have been designed for the Natural History and Science Museum of the University of Porto.

Bente Vinge Pedersen is Head of Exhibitions and has been a part of the team at Medical Museion, University of Copenhagen since 2006, working with every aspect of communication and outreach. She is a historian and is specialised in exhibition curation and production, and has been leading the making of exhibitions in all sizes at the museum. Among these: *Obesity – what's the problem* (2012), *Medicoteknik – Forty years of Danish Biomedical Engineering and Innovation* (2014) and *The Body Collected* (2015).

James Peto is currently on research leave from his role as Head of Programme at Wellcome Collection, London. After three years in the Exhibitions department at the Institute of Contemporary Arts and a year

(1986–1987) on the Whitney Museum's Independent Study Program in New York, James managed public art projects on Tyneside, UK, before becoming Exhibitions Curator at the Whitechapel Art Gallery, London, in 1993. He moved to the Design Museum in 1998, firstly as Curator and later as Head of Exhibitions. In 2005 he joined the Wellcome Trust with responsibility for establishing a programme of temporary exhibitions in the new Wellcome Collection, which opened to the public in 2007. He oversaw an expanding programme of around 35 exhibitions often characterised by an interdisciplinary approach to their subject. He was appointed Wellcome Collection's Head of Programme in August 2015.

Maria Pfeifer holds the position of key curator and researcher at the Ars Electronica Futurelab (Austria), where she focuses on topics of futures studies, art-inspired innovation and art-science-collaboration, and coordinates artist residencies. She studied art, comparative literature and cultural studies in Vienna and completed her master's thesis on the visualisation of literature in 2013. She is also currently completing a university course for cultural mediation. Before working at Ars Electronica Futurelab, Maria was a self-employed cultural manager, researcher, lecturer and content writer. Since 2010 she has also been part of the core team of the *sound:frame Festival for Audiovisual Expressions* in Vienna, specialising on the fostering of intercreative practices and networking activities.

Hannah Redler Hawes is an independent curator, museum professional and writer specialising in projects which bring together contemporary art, science, medicine, engineering and technology, new media and participatory process. She curates exhibitions, events and art interventions with and for a variety of public and private organisations, which have included Science Gallery London at King's College London, FACT Liverpool, Institute of Physics, Tate Modern and the UK Digital Catapult. She has co-directed the Open Data Institute (ODI) Data as Culture art programme since 2014. Between 1998 and 2014 she worked at the Science Museum Group, predominantly as Head of Arts Programme in London. Here she curated interventions, exhibitions and events by emerging, mid-career and leading international artists on the themes of biomedical science, energy, climate science, photography, and digital media. Resultant artworks established the Museum's Contemporary Art Collection. She is a trustee of the Kraszna-Krausz Foundation which promotes excellence in photography and moving image book publishing.

Thomas Söderqvist was Professor in History of Medicine and the Founding Director of Medical Museion, University of Copenhagen in 1999–2016. He was trained in phylogenetics in Stockholm (MSc, 1972) and in philosophy of science in Gothenburg (PhD, 1986). His research and teaching career in Denmark over the last 40 years covers the history of ecology and immunology, the historiography of contemporary science, the genre of scientific biography, and museological problems in sci-tech-med museums. He has published around 100 articles in academic journals and written/edited ten books, including *The Ecologists* (1986), *The Historiography of Contemporary Science and Technology* (1997), *Science as Autobiography* (2003), *The Historiography of Contemporary Science, Technology, and Medicine* (2006) and *The History and Poetics of Scientific Biography* (2007). As Director of Medical Museion for over 15 years, he focussed on building an institution that integrated classical curatorship and basic research in the humanities in co-operation with artists, designers, scientists and clinicians.

Mike Stubbs is the Creative Producer of Doncaster Creates and former Director of FACT, Liverpool (UK), jointly appointed by Liverpool John Moores University, as Professor of Art, Media & Curating in 2007. Previously he was Head of Program for ACMI (Australian Centre of Moving Image), Melbourne & Senior Research Fellow, Visual Research Centre, Dundee University. Encompassing a broad range of arts and media practice, his leadership, curating and advocacy has been internationally acknowledged. Stubbs has commissioned and produced over 350 exhibition programmes including White Noise for ACMI, the Nam June Paik retrospective with Tate Liverpool & Kunst Palace, Dusseldorf, Pipilotti Rist and Sk-Interfaces at FACT as part of European Culture 2008. He co-founded the ROOT (Running Out of Time) and AND (Abandon Normal Devices) festivals and is the co-founder of EMARE, European Media Art Residency Exchange. Curatorial Advisor to Jikji International Festival, Cheoungjiu, South Korea, Digital Revolution, Barbican, London, Advisor, Hyphen hub, New York, Curator Crystalize, Billingsgate, UKTI/KOTRA, London. Mentor, Creative Capital, USA. Stubbs is also an award-winning moving image artist recently commissioned for Next-Art Tainan and has won more than a dozen major international awards, including Barbara London Prize, Echigo Tsumari Trienalle. He is Honorary Professor at Liverpool University and UTS, Sydney. He was made a fellow of the Royal Society for the Arts and awarded a Fleck Fellowship, Banff, Canada in 2002.

Karin Tybjerg is an associate professor at Medical Museion, University of Copenhagen. She did her PhD and a research fellow at Department of History and Philosophy of Science at University of Cambridge and was Head of Modern History and Ethnographic Collections at the National Museum of Denmark. At Medical Museion she has curated *The Body Collected* on medical collections of body parts; *Biohacking – Do-It-Yourself!* an installation of a biohacking lab focusing on co-curating and craft in science; and organised debates on art-science ethics for *Trust me I'm an Artist*. She has written on a range of topics in history of science, on collections of human remains and on exhibiting objects of knowledge.

Nuno Valentim is an architect. He has been in charge of some of the most challenging and enticing rehabilitation projects in the city of Porto – he is responsible for the requalification of Casa Andresen in the Botanical Garden of Porto and its conversion into the Hall of Biodiversity and for the ongoing Bolhão Market Restoration and Modernization project (Porto's main market). He shares his architectural practice, which was recently awarded the IHRU National Rehabilitation Prize 2017, with 12 architects.

Nuno has a degree and PhD in Architecture by the Faculty of Architecture, University of Porto and also a MSc in Building Rehabilitation by the Faculty of Engineering, University of Porto. He is a Professor and Researcher at the Faculty of Architecture, University of Porto.

Louise Whiteley is Associate Professor in medical science communication at Medical Museion, University of Copenhagen and the Program for Metabolic Science in Culture at the Novo Nordisk Foundation Center for Basic Metabolic Research. Her work is split between research, teaching, and curation, and focuses on their combination through experimental public projects. Her research interests include how biomedical approaches to the mind affect how people understand themselves; from neuroimaging studies of mental illness to recent research and commercial interest suggesting that gut microbes affect mental health. A second research strand includes how relationships between art and science operate in collaborative curatorial contexts, and how exhibitions might be used as public research sites. Louise has curated several exhibitions at Medical Museion, including the award-winning "Mind the Gut" (alongside Associate Professor Adam Bencard) which opened in Autumn 2017.

Mark Wright holds a joint UK based appointment between FACT (Foundation for Art and Creative Technology) and the Liverpool School of Art and Design at Liverpool John Moore's University (LJMU), where he is Research Director of FACTLab, a joint venture between FACT and LJMU which embeds research culture, art making and learning inside FACT.

Previously, he has been a senior research fellow between Edinburgh College of Art and the School of Informatics at the University of Edinburgh. He holds a PhD from the University of Cambridge. His research interests centre around the relationship between arts practice, culture and technology and how new meaning unfolds from their interplay through embodied interaction.

Ionat Zurr, PhD, is Chair of the Fine Arts Discipline at the School of Design and the academic Coordinator of SymbioticA, School of Human Sciences, at the University of Western Australia. Zurr is an artist, researcher and curator who co-developed with Oron Catts the Tissue Culture and Art Project in 1996. She is considered a leader in the growing field of

Biological Art, both as a practitioner and a theoretician. Her research was instrumental to the development of SymbioticA in 2000.

Zurr develops and runs the academic program at SymbioticA. She designed and initiated the internationally renowned program Master of Science (Biological Arts) intended for people from all disciplines (arts or sciences) who wish to undertake interdisciplinary studies to engage with the crossover of art and science. Zurr presents her research in forms of exhibitions, workshops and public talks in numerous places around the world. Her work is also present at MoMA NY.

Acknowledgments

We are extremely grateful to all authors, who generously and passionately decided to embark with us on this adventure.

We also wish to thank David Anderson, Thierry Botti, Amiltom Morera, Tom Rockwell and Paola Rodari for supporting us in the identification of interesting case studies and contributors as well as Matteo Merzagora, Simona Casonato, and Daniel Glaser for the inspiring conversations.

A very special thanks goes to Gianmarco and Lorenzo for their love and unflagging support during the many days and nights spent in researching and writing, as well as to our mothers for their perceptive and tender care.

We dedicate this book to Clio, Elia, Ottavia and Leonardo and to their curious and unprejudiced view on the world.

1 Preface

Roger Malina

Cutting down the tree of knowledge: a primer on creating an evolving ecology of knowledge

I am both observer and experimenter in the territory covered in this book, on how the art-science movement of recent decades has worked within and symbiotically with the evolving concept of a science museum as well as the social embedding of science. The Leonardo publications, for whom I serve as Executive Editor, have now documented over 50 years the action of some 15,000 individuals whose work bridges the arts and sciences. I knew and worked with some of the key historical figures mentioned in these pages; Frank Oppenheimer, Richard Gregory, Gyorgy Kepes, Jasia Reichardt, Nicola Triscott among others. I have circulated in the innovative community of practice that has adapted classic ideas of a museum, with those of participatory hands on practices, maker and hacker communities, embedded within a broader social context.

The editors and authors of this volume are an active part of this community and their voices represent useful starting points for future conversations and evolutions.

A new breed of curators

As described in this book the new emerging practices challenge the existing definitions of "art" and "science" museums. At the time that John Dewey and colleagues developed their taxonomy of knowledge, the Dewey Decimal System, Paul Otlet designed his Mundaneum, essentially as a "museum of everything" (Wright, 2014).

Now a new breed of professionals has emerged in the museum world with initiatives such as FACT in Liverpool, the Science Gallery network, but also with artists determined to turn science museums into nodes in a world of ubiquitous learning. With this different framing, various ways of knowing can be confronted such as in the Copenhagen Medical Museum or the SymbioticA labs.

Simultaneously, a new breed of curators emerged to penetrate the art museums hostile to art and technology artists led by pioneers such as Barbara London at the NY MOMA and John Hanhardt at the Whitney.[1] In our Nica article we noted that the kinetic artists of the late 1950s and early 1960s were sometimes told that if they had to plug it in it couldn't be art. Younger curators such as Steve Dietz and Christiane Paul helped change the attitude of some art museums, but ironically some science museums have been more receptive to the digital arts.

C. P. Snow was wrong or at least misinterpreted

Words matter, metaphors matter, taxonomies and ontologies matter. Unfortunately, the "two cultures" metaphor of C. P. Snow (1959) has mistakenly framed the arts and sciences as dichotomies, when in essence they are mutualist systems of thought. The conventional classification systems impose artificial boundaries that are rarely followed by the professionals described in this book.

The value of new ontologies and taxonomies, long known to our networks, is gaining larger recognition. In 2013, the US National Science Foundation funded the study "Steps to an Ecology of Networked Knowledge and Innovation: Enabling New Forms of Collaboration among Sciences, Engineering, Arts, and Design" (Malina *et al.*, 2013). Many of the authors in this book contributed to this report where we emphasised the ecological metaphor of dynamic networked knowledge and continuously evolving taxonomies.

I believe we need to work with both art and science museums to contribute to the necessary changes in perspective that in part are being driven by the disruption of digital culture, as profound as the disruptions caused by the invention of writing or of print.

Cut down the tree; we need forests and ecologies of knowledge that can mutualise in the face of climate change and the other complex problems of our time that require bringing together different ways of knowing. As the sciences of complexity are uncovered, disciplinary and reductive scientific methods are often incapable of addressing problems that are intractably complex. Complexity Science introduces different notions of causality that lead to emerging phenomena. This book charts the beginning of what will be a multi-century process.

In-reach as well as out-reach: redesigning science itself

My own career has been in astronomy, building and using space telescopes beginning in 1968 at MIT, and my last position was as Director of the Astronomical Observatory of Marseille Provence. During these 60 years I have participated in and witnessed a dramatic restructuring of astronomy as a discipline studying the universe beyond the earth's atmosphere. The innocent move from photographic film to digital technologies has triggered unexpected epistemological consequences that are currently in their stone age.

Digital astronomy has enabled observations from orbiting observatories, opening up the whole electromagnetic spectrum, and now neutrino and gravitational astronomy are at work from the ground. In the words of Carlo Rovelli (2017), ‘Reality is not what it seems’. We now understand better that our human senses have often misled our ideas on the content and evolution of the universe, and a result biased much scientific research of the past. A number of science historians such as Peter Galison and Patrick McCray have been deepening many of these discussions.

I argue that one of the roles of artists today is making science “intimate”, i.e. translating things that our senses are blind to into a form that can be engaged by our senses. Inevitably the work of artists today will change the history of science in the future. Science museums that have championed these artists, sometimes as simple tools of science communication or “outreach”, have helped a process of “in-reach” where artists are changing science itself.

Many of the projects described in this book help make “science intimate” in a variety of sometimes provocative approaches as with SymbioticA. Bioartists are engaged deeply in questioning and exploring the implications of genetic engineering and the new biological technologies including artificial life. These groups engage in art as a critical practice but go beyond critique to drive scientific ideas and methods in new directions. The EU funded “Trust me I am an Artist” project directed by artist Anna Dumitriu and ethicist Bobbie Farsides under the Waag Society’s leadership is one example that I have been involved in with Annick Bureaud and the French Association Leonardo.

In my past, I have also worked for a year as an Osher Fellow at the Exploratorium, an experience which led me to understand that the scientists’ obsession with “outreach” was myopic. Science needed to synergise more deeply with people’s interests, and our museums meet the non-scientists’ desire to “in-reach” and redirect science in new directions. Helga Nowotny (2003), former president of the European Research Council, called for a ‘socially robust science’. Her implication was that today’s science, far from the endless frontier helping us win the peace as Vannevar Bush (1945) argued, had become socially ‘fragile’. Scientific priorities were usually set within a narrow community of experts, framed by the funding objectives of governments, corporate interests and political expediency.

A different perspective is provided by G. Rattray Taylor (1953), who notes that moral attitudes on sex in the middle and upper classes have swung back and forth, correlated also with periods of “patrism” and “matrism”, but more importantly in the context of this book, with periods of intellectual exploration, such as the Renaissance, outside existing taxonomies of knowledge. It is perhaps no coincidence that some of the events described in the book occurred during the political and sexual turmoil of 1960s, whether it be Jasiah Reichardt's seminal exhibit Cybernetics Serendipity, Frank Oppenheimer's founding of the Exploratorium, E.A.T. and numerous other events described in this book from Japan to the Balkans to Latin America.

As detailed in this book, the curators and artists struggling to incorporate the work of the emerging "art, science, technology" professionals in science museums is exciting and difficult. Large institutions, such as science museums, inevitably reflect the world view of their funders and because of their size respond slowly to external societal changes. Coupled with the disruptions being triggered by a "digital culture" as argued above, the very systems that are used to determine the content and activity of museums are shifting and curators and artists find themselves to be risk takers pioneering new ontologies and taxonomies. The act of curating then becomes one of the mechanisms for redesigning science itself.

The emergence of big data and artificial intelligence has also influenced the scientific method as digital technologies that have enabled us to better understand how our senses have biased the scientific method itself. Digital cameras provoked the "virtual observatory" movement allowing astronomers to study hundreds of millions of objects at many wavelengths simultaneously. Artificial intelligence algorithms are now used to find patterns in data, unbiased by the patterns humans usually seek, such as symmetry or other human sensory (or intellectual) morphological lenses. These algorithms cannot describe how they made the discovery, or what other hypotheses they used, which is something that is usually required in the scientific method. Yet supercomputers are now used to "simulate" the universe, or global climate change, and this matching of a simulation with the data becomes a form of hypothesis making and testing. Needless to say, 18th- and 19th-century science would not have recognised these methods. As described in this book, information artists and data artists have invaded the territory of data collection, coming up with techniques that had not been imagined by the scientists.

A role for science museums in contributing to redesigning science

In an 1895 article, Arrhenius argued that variations in trace constituents – namely carbon dioxide – of the atmosphere could greatly influence the heat budget of the Earth (Arrhenius, 1897). He underestimated the rate of production of CO_2 and overestimated the impact on global temperature, but in all the fundamental aspects of the modelling he has been proved right. Yet perhaps it is no accident that these results have taken a century and a half to convince certain heads of state in our contemporary world. Perhaps the way science museums were developed over the past centuries, particularly after science was restructured after World War I, are part of the problem. To be somewhat provocative: what would have happened if, beginning in 1895, science museums had taken on explaining that the accelerating use of carbon-based fuels would have catastrophic consequences. Artists of the time could have translated the scientific data, on a molecule that our senses cannot detect, into a form that could be experienced. The literature on embodied knowledge reinforces these ideas.

Science museums, with the science academies and other scientific institutions including artists, could perhaps have helped convince governments that their policies could mitigate global warming, so that it would not endanger the sustainability of the societal model that was being developed. Different artists using smart textiles have pioneered this idea with smart textiles that change colour depending on the level of certain chemicals that our senses cannot detect directly, but 100 years too late. More recently artist Dawn Faelnar has been working on an "Earth-Suit" that allows the wearer to sense the weather on other planets than the earth. Could this artwork contribute to art or science or both? At the very least artists might contribute to developing terminologies to describe atmospheric phenomena on other planets that do not occur on Earth.

Concluding thoughts

The very nature of the work of the authors of this book is un-disciplinary and the idea of an "art-science" museum would be a contradiction in terms. We need to find ways to cut down the tree of knowledge and replace it with an evolving network of knowledge. Unfortunately, our institutions are mostly structured in static 'tree' structures, as are disciplines in universities. Disciplines are important tools for certain kinds of problems, but we should beware of mistaking the tool for the purpose. The path might be hazy, yet the only way to proceed is with a spirit of curiosity and experimentation, which this book encourages from the first page to the last.

Note

1 Isabel Sena has recently synthesised the history of these pioneering new kinds of curators in her online course "Curating in Art, Science & Technology: Potentials and Pitfalls" (July 2018).

Bibliography

Arrhenius, S. & Holden, E.S. (1897) On the Influence of Carbonic Acid in the Air Upon the Temperature of the Ground, *Publications of the Astronomical Society of the Pacific*, 9(54), 14–24.

Bush, V. (1945) *Science the Endless Frontier*, Washington, DC: United States Government Printing Office.

Malina, R.F., Strohecker, C., LaFayette, C. & on behalf of SEAD network contributors. (2013) *Steps to an Ecology of Networked Knowledge and Innovation Enabling New Forms of Collaboration Among Sciences, Engineering, Arts, and Design*, National Science Foundation [Online]. Available at: https://seadnetwork.wordpress.com/report-and-meta-analysis/ (Accessed 03 July 2018).

Nowotny, H. (2003) Democratising Expertise and Socially Robust Knowledge, *Science and Public Policy*, 30(3), 151–156.

Rattray Taylor, G. (1953) *Sex in History*, London: Thames and Hudson.

Rovelli, C. (2017) *Reality Is Not What It Seems: The Journey to Quantum Gravity*, New York: Riverhead Books.

Snow, C.P. (1959) *The Two Cultures and the Scientific Revolution*, Cambridge: Cambridge University Press.

Wright, A. (2014) *Cataloging the World: Paul Otlet and the Birth of the Information Age*, Oxford: Oxford University Press.

2 Introduction

Camilla Rossi-Linnemann and Giulia de Martini

We indeed live in a world of fierce contradictions and clashing opposites.

Science research and technological developments have reached unprecedented levels of refinement, opening up extraordinary as well as intimidating scenarios.

Western democracy has incubated the seeds of equality and independent thought, growing at the same time the healthy branches of relativism and those of a post-truth society where facts seem to succumb to emotions and personal beliefs. In today's global society, values are explicitly shifting and subject to constant negotiation.

Where do we, as science museum practitioners, stand in this picture? Which role do our institutions play? And what has art got to do with all of this?

We argue that there is a growing number of situations in which cultural institutions dedicated to science engagement[1] have opened up their doors to art for a number of different reasons and in different ways. We intend to present this scenario, not through an analytic mapping of such situations but rather through a qualitative investigation that brings together practitioners and thinkers in the field to reflect openly on how they are doing what they are doing and why.

Moreover, the intention is not to build a strong case for arts in science centre/museum settings,[2] but rather to understand what specific role art can play in these contexts: if its task is limited to being a "tool" of science communication; if it brings threats or benefits to fact-based education of traditional science museums; if it actually ends up somehow challenging the very nature of these institutions and reframing their mission and position in society.

In curating this book we were moved by different contrasting questions: can arts be considered a tool for broadening the range of visitors and democratising information, or might they push away specific categories of audience and staff? Can they serve as a tool for clarifying scientific/technological content, or do they add an extra layer of complexity to messages conveyed in science communication venues, mingling the reality of hard data with imaginative scenarios? Can they help curators offer new integrated knowledge perspectives, or might they create schizophrenic narratives where scientific and artistic contributions coexist as separate elements? Can they be used to involve audiences more inclusively, or might they disruptively endanger traditional curatorial hierarchies? Can they bring useful ethic-related perspectives to hard science, or might they foster purely emotional responses that ultimately interfere with the development of critical fact-based and scientific thinking?

Art and science combinations: an expanding ecosystem

The strict ground limitation set to this book has helped frame the area of work and analysis, yet what is happening in the field of science centres, museums and similar institutions is of course part of a broader frame of events.

If we take one step back and look at today's cultural panorama we see examples of cross-pollination between artistic, humanistic, scientific and technological work at all levels. Strong forms of interdisciplinary practice have developed spontaneously in society also as a result of historical, economic and political conditions (where a relatively long-standing peace in part of the globe has allowed for travel, dialogue, cultural thinking, and collaboration on global issues) and the development of information technology (that has enabled, among other things, bottom-up content production and knowledge sharing economies).

Spontaneous interdisciplinary communities are thriving thanks to online platforms, sometimes creating physical spaces managed by representatives of maker, hacker, DIY (Do It Yourself) and DIWO (Do It With Others) movements, that base their very existence on the sharing of expertise and competences, and that lead to new concepts such as Open Innovation, Co-design, etc.

Another area in which interdisciplinarity is being openly discussed is the world of formal education, with its various stakeholders, from universities to curriculum developers.

At academic level, interdisciplinary research is often a reality and academic faculties are struggling to keep up with the naming of new areas of specialisations. Biophysics, bioinformatics, digital humanities, social design, visual mathematics are only a few examples of newly born "disciplines", which translate into programmes, schools, institutes, departments and courses. Alongside these areas of knowledge that have already been renamed there is a growing number of university programmes that foster more free interdisciplinary experiments bringing together students and researchers from different areas of knowledge. The underlying idea is that sustaining the convergence of approaches and people is ultimately key to the necessity of solving creatively the global issues that society faces today: from environmental to health, food and safety challenges.

Born in the US to counterbalance national learning programmes (such as the "No Child Left Behind" programme or Obama's "Race to the Top" framework), the STEAM[3] movement is advocating for the inclusion of Arts in Science, Technology, Engineering and Mathematics curricula. The term STEAM does not actually refer to a definite and fixed set of ideas, yet it is used loosely by all those who somehow promote the necessity of breaking down strict disciplinary boundaries in schools. STEAM supporters maintain that broader approaches—which integrate design thinking, problem-solving and enquiry-based methodologies—are ultimately more useful in science education itself (McDougall *et al.*, 2012) and that they can be associated with improved academic performance (Gazzaniga *et al.*, 2008). Edwards (2008) attempts to describe the "catalyst" elements of creativity and innovation, which he claims lie beyond theory and can be identified in a mix of science and art based experiences.

Alongside study reports connected to specific curriculum development projects (Informal Science, 2017) there is mostly grey literature on this subject and the scenario is still very contrasted (Xanthoudaki, 2017), yet many, including Robert and Michele Root-Bernstein (2001), recognise that arts and crafts are activity spaces where critical thinking skills can be successfully developed alongside other 21st-century competences such as problem solving, creativity, and communication and collaboration skills.

The idea that interdisciplinarity is at the base of a healthy, creative and productive society is also reflected in much public narrative and in policy making statements such as the "New Strategy for American Innovation" promoted by former President Obama in the US (National Economic Council and Office of Science and Technology Policy, 2015) or the moving from the concept of a "knowledge-based society" to that of "innovation-based society" promoted at the European level. The pressure for industrial innovation and economic growth is in fact palpable in many national agendas and cuts across all areas of public policies, from that of education, knowledge production, and economic development.

In the final report of the European Research Area Board titled *The new Renaissance: will it happen? Innovating Europe out of the crisis*, 2012 we read: 'The current buzz on social innovation is a wake-up call on the role of social sciences and the humanities. New, more efficient local and global networks are changing the world and transforming traditional approaches to research policy' (European Commission, 2012).

Economic, legal and operational difficulties still hinder collaborations, yet a number of funding programmes have somewhat sustained political agendas. Furthermore, large national funding bodies in the US and UK,[4] as well as smaller foundations, have launched support programmes such as the INSPIRE and AHRC hubs encouraging the formation of incubators contributing to the support of a "creative economy" (Fox Harrell & Veeragoudar Harrell, 2010; Glinkowski & Bamford, 2009).

The "Space, Ecology, Art and Design network" report[5] states that 'since the 2000s, successful collaboration among scientists and engineers on the one hand, and artists, designers, and the humanities on the other, has been accelerating [and that] this has generated emerging practices that impact work and have potential to mitigate the difficult problems of our times' (Malina *et al.*, 2013, p. 1). A paradigm shift seems to be taking place in newly forming knowledge-communities: these are outcome-focused (rather than being based on a disciplinary corpus) and they seem to seek interdisciplinary collaboration spontaneously.

Whilst any study referring to the relationship between art and science cannot be conducted without mentioning Snow's reference to two separate cultures within society (Snow, 1959), we argue that this division must be put into historical perspective, as it is itself a product of its own time. In fact, we must keep in mind that in the late 1950s, when Snow first used his unpredictably long-lasting expression "the two cultures", public opinion was still sensitive to the dark shadows cast upon the scientific community by the shock of nuclear weapons and of World War II. Snow was advocating for a dignified place for science in the intellectuals' community and in the arena of high-culture, still grandly ruled by the humanities.

If anything, today it seems like it is the field of humanities that feels the need to battle for a place in a society which is seen mostly as driven and ruled by science and technology. As funding for culture and research becomes scarce, humanistic research is often drawn to join science-based projects that seek to produce concrete innovation outputs (Sullivan, 2005). Yet this pressure on goal-oriented and impactful studies represents both a challenge and an opportunity for research in the arts, humanities and sciences alike.

Alongside situations in which art and science practitioners work effectively together to solve practical problems, there is a deeper level of collaboration that can be fostered, that of comparing and cross-breeding research methodologies. This is still often seen as an impossible and epistemologically wrong attempt, as art research and science research are said to be based on different paradigms.

In contrast, Wilson (Wilson *et al.*, 2013, p. 45) argues that scientists and artists not only share the commonly mentioned inclination towards curiosity and imagination, but also work similarly in terms of deep-thinking patterns. If research's primary focus resides in 'questioning and challenging fundamental conceptual structures, models and metaphors in one way or another, or seeking to propose a conceptual model where none has been previously acknowledged', then both artists and scientists share a Bayesian approach to reality, which implies that we make sense of the outside world through our senses and then formulate provisional theories and constructs (Wilson *et al.*, 2012). This entails that researchers in art and in science share, to a certain degree, a common sense of uncertainty towards their interpretations, which entails the possibility to look at one another's disciplines to cross-fertilise methodologies and epistemologies that would otherwise never be questioned within one's own "research tribe".

Many large research centres such as CERN, ESO, JRC, NASA have launched their own artist collaboration programmes hosting artist residencies in their facilities or collaborating with artistic platforms.[6] Yet the aim of these programmes is often linked to matters of communication and they can frequently become "one way" situations as – unless particularly open individuals are involved – they are designed to provide artists with new inspirations rather than to affect the work of science researchers itself. In other programmes[7] scientists and artists are treated more "equally" with environments that encourage free research, more disconnected from set objectives.

The interest of artists in science might be more easy to understand: in the last century, artists have openly turned to science for the richness of its subject matter as well as for the new expressive tools made available by evolving technologies (x-rays, holography, imaging techniques, etc.). Sciart, Bioart, Technoart, Nanoart are all new denominations of artistic currents that loosely connect to different areas of science.[8] Artist-physicist Arthur Miller (2015) has identified the emergence of a "new art movement" that developed precisely from the merging of art and science, highlighting the influence that contemporary scientific research has on contemporary artists.

Yet what can scientists gain from art? Zilberg (2015) issues a note of caution on the power of art to contribute to scientific discoveries. The impact of art is still uncertain or, at best, unmeasurable. Yet many collaborations, from the 1930s onwards have proven effective in terms of scientific results: from Edgerton's introduction of film-making, strobe light and time-lapse photography for the investigation of physical and environmental phenomena to today's artificial intelligence developments investigating the very "structure" of creativity through artistic research such as music composition (Dabby, 2008).

A spontaneous reconnection between different academically defined fields of knowledge is taking place, and today's science centres and museums are more or less consciously shaped by this changing ecosystem.

From Renaissance collections to new generation hybrid museums

The idea of combining art with science communication is all but new. In his chapter in this volume, Pietro Greco argues that since the dawn of time art has been powerfully shaping our collective image of science. One needs only to think about the influence of Vesalius' and Leonardo da Vinci's drawings on the surrounding community of intellectuals and on the Renaissance equivalent of today's researchers to understand how artistically refined tools could intertwine deeply with scientific and technical studies.

In the 1500s, a number of wealthy intellectuals began to collect natural objects. The recent discovery of new species and objects in "America" had in fact brought to Europe fascinating items that created problems in the existing classification systems, encouraging collectors to propose new taxonomies and displays, such as the baroque arrangements of Ulisse Aldrovandi's collections. "Theatres of Nature" and "Wunderkammer" were created 'to understand and dominate reality [. . .]. As places of experimentation, where objects were not only contemplated, but also manipulated, researched, measured, compared' (Olmi, 1997, pp. 48–74, translated by eds.).

During the 1600s, new places were constructed to display collections: scientific academies and the first major public museums were born, such as the Jardin Royal des Plantes Médicinales – the first nucleus of the National Museum of Natural History in Paris – founded by Louis XIII in 1635.

These collections – whose principal beneficiaries were research scholars – included objects of research and they were often displayed with little order until, in the 1800s Linnaeus' classification system brought order and clarity into science museum displays. 'The museum of natural sciences, for the first and only time in history, was set up so that the articulation of its spaces mirrored perfectly the scientific system' (Merzagora & Rodari, 2007). The progressive affirmation of the different scientific disciplines was paralleled by a multiplication of the museums dedicated to them: paintings ended up in art museums, furniture in history museums, seashells in natural history museums, etc. The products of the new industrial revolution were collected for the benefit of universities and industries in places such as the Conservatoire National des Arts et Métiers in Paris (1794) or the Franklin Institute in Philadelphia (1824), which used collections to train new craftsmen and designers.

By the end of the 19th century the debate on the organisation and classification of nature had reopened, leading to a discussion on exhibition methods as curators faced a new problem: they wanted to represent not only objects, but also phenomena, theories, and stories such as Darwin's evolutionary theory or Maxwell's electromagnetism. The number of objects collected decreased and, as a consequence, researchers no longer saw the museum as a natural habitat for research. Museums gradually closed their doors to "those who knew" and opened them to "those who did not know".

This new generation of science museums was represented by places such as the Deutsches Museum in Munich (1906) or the Museum of Science and Industry in Chicago (1933). These institutions became public instruments for bringing culture and knowledge to the large public and for supporting national approaches to the popularisation of science, as was done throughout Europe in the 20th century (Schirrmacher, 2013). Display techniques in these museums were privileging narrative rather than classificatory principles, yet preserving and collecting were still a core part of the institutional goals.

World Fairs and the need for improved public understanding gave birth to many more national collection-based museums (Canadelli *et al.*, 2019) and to a third generation of museums, now known as science-technology centres (Friedman, 2010), where public education is the only goal, and Fair-inspired animation and interactivity are adopted as primary display and educational techniques.

The Exploratorium in San Francisco, founded in 1969, was the purest example of this new generation of institutions, where there are no historical collections, where exhibits focus on phenomena, and where the general audience's experience is the shaping force of the display. The emergence of such places was the result of many factors and of the spontaneous meeting of different worlds. The history of the Exploratorium itself, well described by Marina McDougall in her chapter in this volume, is the result of the inextricable entanglement between the lives of physicist Frank Oppenheimer (founder of the Exploratorium), art curator Jasia Reichardt, neuropsychologist Richard Gregory (founder of the Bristol Exploratory in 1987), art historian Ernst Gombrich and many others.

In those same years, the art world was opening up to experimental spaces and situations that connected artists with emerging technologies. Electrical engineer Billy Klüver[9] was one of the founders of the E.A.T. (Experiments in Art and Technology) in 1966. Two years later, MIT's Center for Advanced Visual Studies (CAVS) was founded by the Hungarian born artist György Kepes, and in 1979 the Ars Electronica festival in Austria was conceived, to then incorporate a fixed research laboratory and a permanent exhibition space in 1995.

Influenced by the science centre model and by contemporary art research, a number of "hybrid" institutions were born among which we cite in this volume: the Foundations for Art and Creative Technology – FACT in Liverpool (2003); the Wellcome Collection in London (2007) rooted in a Sci-Art programme initiated in 1996; the Science Gallery in Dublin (2008). It would be impossible to name all new-born establishments, but it is worth noting how – somewhat echoing first generation museums such as the Palais de la Decouverte – many of these institutions have adopted more wide-ranging designations such as "Pavilion of Knowledge" in Portugal, "Quai des Savoirs" in France and so forth. This attitude seems to reflect a tendency to transcend disciplinary boundaries, viewing science museums as places that treat reality in its multifaceted complexity.

Exploring how art is influencing Science Museums

Adopting the definition used by the curators of the Russian Pavilion at the Venice Biennale 2019,[10] the museum is 'a kind of organism living by its own laws and capable of creative endeavours beyond the ability of any of its component parts – objects, people, or buildings.'

We argue that art is now brought into the science engagement context in different ways: (1) as a narrative tool; (2) as an ally to sustain education; and (3) as a disruptive element that sparks discussion, imagination and triggers self-reflection. Let us introduce these three concepts that will then be further explored and exemplified by the chapters collected in this volume.

Art as a narrative tool: seeing the unseen

One way in which artists are commonly brought into the curatorial picture of science museums is in the role of "translators", i.e. professionals who can convert into aesthetically engaging (and often synthetic) forms a set of data. In this case artists can be asked to produce works that are in fact data-visualisation products. Yet we must not think of data visualisations merely as graphics. Creations can take multiple forms and result in beautiful installations that organise objects or information into attractive museographic compositions that convey complex layers of meaning. An example is provided by the *Hypercubic Showcases of Sudden Comprehension* at the Hall of Biodiversity in Porto (Portugal) described in this book. One of these display cases shows hundreds of eggs of different size, colour and shape arranged into an elegant cubic composition which is both extremely pleasing to the eye and which also triggers spontaneous questions regarding the extreme variety and subtle variations of nature and evolution. Such installations, which we associate to pure data visualisation, are becoming common also outside museum walls, in the realm of public art, where artists are using their works in association with activist movements to produce works such as Andrea Polli's *Particle Falls*, a colossal projection that visualises current levels of fine particulates in the air cascading down the facades of iconic building.

When science curators bring artist in at an early stage of their research and of the construction of the exhibition concept, the results are interesting hybrids where both the content and the form of the displays are defined by a plurality of views. An extremely interesting example of this interwoven design is represented by the displays at the Medical Museion in Copenhagen (Denmark) (Ken Arnold *et al.*, in this volume), where the curatorial team worked closely with artists to solve the issue of displaying the "invisible" (concepts and objects) of biomedicine.

What happens in fact when a museum curator wishes to present something that is real but either too dull to the eye[11] or which cannot be grasped directly by our senses? How do we display a real phenomenon (or object) that is too small (for example a cell or a nanotube), too large (for example a galaxy or all existing YouTube videos), too slow (for example the growth of a plant), too fast (for example the passing of a bullet), too opaque (for example the inside of a body), too far (for example the bottom of the sea), too dangerous (for example the radioactive decay of uranium or the biting of a shark) or so massively distributed as to transcend spatiotemporal specificity (such as Timothy Morton's (2013) so-called hyperobjects, in eg. global warming, Styrofoam, and radioactive plutonium)?

Art can indeed come to aid by producing powerful metaphors. A.B. Byatt (2000) defines art as a natural expression of the human mind in the sense that, just like 'metaphor-making [it] is a fundamental part of the way the human mind makes connections. [. . .] art explores connections in ways very different from science's orderings'.

If metaphors are powerful narrative tools, so are objects themselves, and the intervention of artists can sometimes help identify those objects that have an intrinsic aesthetic power, such as natural specimens or technical devices such as fog chambers (which were chosen both by Oppenheimer for the Exploratorium, and by artist Tómas Saraceno for his recent exhibition "Gravity"[12]).

In his idea of a "total museum", Jorge Wagensberg (2005) advocated for the importance of presenting real objects (made of matter that takes up space) and phenomena (changes experienced by objects, taking up time) through all museological means available, including those museographical techniques "stolen" from contemporary art exhibitions.

Similar considerations about a certain intrinsic beauty can apply to scientific images. There are many artists who choose to work directly with scientific images created through scientific imaging devices and techniques, from early research with telescopes and stereoscopy to Seiko Mikami's VR explorations of molecular structures, Eduardo Kac's works with luminescent transgenic bacteria, Drew Berry's molecular animations.

The collaboration between artists and visual sciences goes way back in time and has created interesting synergies also from the point of view of scientific results. For example, today, digital interfaces and the visualisation of big data are a crucial part of research processes, as they actively inform the way data is interpreted and treated. The idea that artists are not only creating products but also working on software and interfaces is successfully exemplified by the Mathenaeum exhibit created for the Momath in New York (USA).

In this volume, Hanna Redler Hawes describes the experience of introducing contemporary art in the Wellcome Wing of the Science Museum in London (UK) with the hope it 'might offer ways of prising open "gaps" in its narratives, encouraging visitors to challenge the assumptions of science and ask their own questions'. As she points out, the pure introduction of artworks in an otherwise completely "scientifically curated" environment posed questions of relevance and expectation, which ultimately brings us to reflect on how the process of bringing artists into science engagement institutions can result in more sterile results or in more deeply changing experiences depending on the level of involvement and conviction of the whole museum ecosystem.

Sharon Macdonald and Paul Basu (2007) trace an interesting parallel between experiments and exhibitions. They argue that just as the purpose of a scientific apparatus is "to make visible the invisible", the purpose of an exhibition is also that of showing, displaying. Yet this process of displaying in contemporary museological practice should not be thought of as a pure act of reproduction, but rather as an act of experimentation, during which new knowledge is produced. If we adopt this point of view, the inclusion of new visualisation media in a curatorial narrative is all but a purely "descriptive" act. Introducing arts in scientific curation can be an act by which new and previously inexistent knowledge is produced.

The opportunities represented by the inclusion in exhibitions of new non-textual media (photography, painting, installations, videos, etc.) are infinite and the idea of drawing from different disciplinary traditions should be an active element of museological experimentation itself.

Art for science education and enquiry: patterns of thinking

As Nina Simons puts it (2013), 'in science centers, we try to combat the notion that science is complex work for a limited, rarefied few. So we focus on the idea that "you can be a scientist" and that "science is fun". Do these democratising messages prevent us from pursuing interesting ways to present the extraordinary genius of some scientists and the incredible complexity and repetition of scientific work? In art museums, we try to combat the notion that art is something your child can do, and if you like it, it is art. So we focus on the idea that "artists are special" and that "art is complicated." Do these elitist messages prevent us from exploring useful ways to honour the creativity in everyone and the simple pleasures of aesthetics?'

Of course this latter concern does not relate to all art. Quite the contrary. In the 1960s, Public Art reacted strongly to the elitist art market where art production (and fruition) were a realm of the few. Numerous artists embraced the idea of an art practice based on inclusion. This plethora of experiences gave birth to different artistic strands, where communication was more important than celebration. Site specific installations escaped the formal walls of art-dedicated institutions to conquer the streets whilst forms of performances and participatory art invited the audiences to be protagonists of the creative process itself.

It was in this climate that artists and science communicators, science museum curators and educators found a fertile common ground for collaboration.

At a very basic level, science educators and communicators may turn to art for its aesthetic qualities. Humans seem to be naturally attracted to beauty and harmony and visitors entering a science museum are no doubt immediately attracted by "beautiful" and "strange" displays and exhibits. Thus, as mentioned above, museographers will often seek the help of artists to create something that is spontaneously attractive, which sparks curiosity and interest, and draws the viewer's attention. At the basis of this strategy is the idea that aesthetics plays a crucial role in the learning process.

The notion of "beauty" is complex. Yet – among all other things – beauty seems to represent a privileged channel of communication between our inner self and the external world. Simplifying to the extreme: we tend to classify as "beautiful" something that makes us feel good, that we somehow understand, that has an internal harmony, and which ultimately does not distress our mind and senses. We seem to be naturally attracted to things and concepts which we perceive as aesthetically engaging, functional and useful.

Cognitive psychology (Page, 2000) shows us that our brain works in a very specific way: when faced with something new or unknown, it activates a sophisticated recognition process which allows us to evoke previously memorised items that somehow associate with the new object/concept, recalling related information (sensory, semantic, but also emotional).

So on the one side, we have our spontaneous attraction to beauty, and on the other the fact that we tend to learn better when we manage to relate unfamiliar objects to areas of familiar pre-knowledge. This entails that art which is aesthetically reassuring, or which somehow taps into the pre-knowledge of the viewer/perceiver, can more easily be introduced in educational settings.

A practical result of this collaborative educational scenario is represented by the development of "interactive exhibits", such as those initially conceived at the Exploratorium in San Francisco and now present in many other science centres and venues.

Most classic Exploratorium exhibits "present" a physical phenomenon leaving a significant degree of freedom in the user's personal exploration and/or observation process. Air cannons, fog tornadoes, laser tables, and chaotic pendulums are all common exhibits characterised by what Allen and Gutwill (2004) call PAR features: "Physical" (not involving symbolic or computer-based media); "Adjustable" (meaning that the exhibit can be varied in a way that is continuously changing, beyond a simple "on/off" push of a button); "Relevant" (in which interactivity supports direct exploration of the phenomenon being exhibited, rather than being used to entice visitors to read informational text or graphics).

Artists have played a fundamental role in the development of these classical exhibits, creating devices that sensitively and respectfully revealed the phenomena themselves. Environmental artist Ned Kahn, who collaborated for many years with the Exploratorium, declared: 'I strive to create artworks that enable viewers to observe and interact with natural processes. I am less interested in creating an alternative reality than I am in capturing, through my art, the mysteriousness of the world around us' (Kahn, 2010).

The "Center for Art & Inquiry" and the results of the 2011 Exploratorium *Art as a way of knowing* conference described in this volume by Marina McDougall investigate precisely how enquiry processes which are commonly enacted by artists can be studied and incorporated in museum practices to become powerful tools of learning.

As Oppenheimer points out (1979), art and science research are both considered as lenses through which we humans look at the world, investigate it, describe it in terms of patterns and consequently understand it and devise new and previously non-existing interpretations of objects/concepts. 'The works of artists and the didactic demonstrations of scientists and engineers combine to do more than show the sights. They alter, each in a characteristic mode, the way in which individuals perceive both their past and future experiences, and they make people aware of aspects of their surroundings that they have either learned to ignore or never been shown how to see' (Oppenheimer, 1972).

"Participatory art" and artist-led collective artworks show that, on many occasions, the boundaries between art done by artists and art done by audiences may not be so rigid, and that there is an empowering component in these forms of art that somehow mirrors the practice of many science museums to put the visitor in the shoes of scientists, inviting them to become part of scientific research processes through citizen science experiences, living labs and similar programmes.

Such practices are deeply inspired by the work of psychologist John Dewey (1934), who maintained that what is important in the art (and learning) process is the experience of the process itself rather than the physical output. This shift also entails that a fundamental characteristic of effective education is the fact that the learner must experience first-hand a process that will affect him/her personally, deeply and lastingly, thus placing educational experiences at the base of individual personality and of society.

The qualities of open-endedness which characterise artistic research are applied by science educators not only to exhibits but also and foremost to activity and workshop design. Tinkering methodologies, first developed by the Exploratorium Tinkering Studio (rooted in research conducted at the Massachusetts Institute of Technology in Boston) and recently adopted worldwide, rely on a balanced mix of artistic creativity, individually led learning processes, engineering skills and technical rigour. The result is an array of activities that invite individual participants to engage on a very personal level, creating objects and following enquiry patterns that are truly meaningful to them precisely because they incorporate an emotional and aesthetic ingredient. The products resulting from tinkering activities are not solutions to a given problem, they are the expression of personal findings based on personal needs and views. They are unique, original and unrepeatable. In the process, participants acquire scientific and technical skills, but their creations have an aesthetical, ethical and psychological value that closely relates to the nature of art itself.

At individual level, the design process of art-science activities described in this book by Samar D. Kirresh, is rooted in the idea that social values and personality (the Self) are two elements that can no longer be ignored when facing scientific discourses and that the creative and artistic component of art-science activities helps bring out these aspects throughout the learning experience. When art is brought in as a means of self-expression for visitors (as happens in making, hacking and tinkering activities) it opens up a dimension of freedom that brings with it an empowering message: each individual vision/idea is valuable and we are all ultimately able (and allowed) to express our ideas as independent thinkers, to guide our own learning processes, to become confident creators.

Curators and educators at Ars Electronica in Linz also highlight in their chapter how the engagement of visitors in critical discourse can be initiated using four different strategies: artistic intervention, artist collaboration, future ideation and speculative intent, all of which are based on the idea that museums should be "discursive venues" where visitors can feel that they can explore their personal values, imagine society and think about the future of humanity.

When this attitude is brought to its extreme, the science centre/museum might even be transformed in its very nature. As visitors are more deeply engaged and somehow feel equal in their ability to create content, institutions may shift from being communication tools (where a more restricted group of "professionals" finds ways to educate/engage a wide group perceived as non-informed) to participatory creative platforms.

As Bandelli and Ian Brunswick describe in this volume, art is not so much a tool to source visitors' ideas and needs directly. It is rather a tool for curators to build an empathic and ethnographic approach that helps them create a successful exhibition context in which visitors have already played an active role and to which they instinctively respond, ultimately facilitating engagement with a subject that might otherwise be so broad or complex that it can paralyze active thinking.

In living labs and similar spaces, the encounter of artists, innovators, scientists and the public is producing new ideas and even new products. Merzagora (2017) maintains that 'future challenges within science communication lie in a "grey area" where the frontiers between production and sharing of knowledge are blurred. [. . .] Settings are emerging, where we can provide real contribution to scientific research and at the same time facilitate the publics in their process of hacking scientific knowledge to serve autonomously defined and often unpredictable functions'.

Art, inclusion, controversy and imagination: from facts to values

As discussed above, art can help bring to the scientific discourse fresh and personal perspectives that come from non-experts, incorporating insights that could not otherwise be shared and, even more importantly, implicitly asserting that certain science topics, such as biomedicine or artificial intelligence, are 'a legitimate topic for cultural activity'.

As emerges from the Berlin 2017 *Contemporary Art in Natural History Museums and Collections* Conference 'in a growing number of organizations around the world, stimuli are being sought from the arts to try out fresh ways of framing and responding to issues. The underlying assumption is that bringing people, processes, and products from the "foreign culture" of the arts into the workplace helps to stimulate new ways of thinking and acting by irritating routines, challenging established mindsets, and developing new skills'.[13]

So artworks are very often included in science exhibitions with the aim to stimulate discussion about future scenarios that might result from today's developments of science and technologies and from the ethical, political and economic choices that are made in connection to these developments.

Artists' points of view and interpretations can be very effective to underline, emphasise, make visible the problematic aspects of scientific applications or technologies as highlighted in this volume by the work of artists Oron Catts and Ionat Zurr.

It must be acknowledged that in their process of interpretation of reality, artists may include an aspect of distortion and imagination which is part of the specific inquisitive process of art research (Wagensberg, 2017). Many authors in this volume have recognised the role played by imagination in the frame of science engagement. In particular, Mike Stubbs and Mark Wright underline how our relationship with uncertainty and our compulsion to imagine worlds beyond known evidence extends our philosophical understanding and perception of the world.

Of course, this ability of artists to provoke strong emotional responses brings with it great opportunities and great risks for curators, educators and exhibition designers who are ultimately trying to foster critical thinking and understanding rather than emotionally confrontational attitudes.

Curators must be highly aware of the intrinsic power of visionary imagination and of the need to make audiences conscious of the intellectual frame they are invited to move in: at some point there must be a clear, honest and transparent distinction between facts, interpretations of facts, and imagination. So if visitors walk through a science museum's door expecting facts, it must be clear that they will also be offered an open critical and philosophical frame in which to place them.

For science engagement institutions, this entails an extremely profound shift in attitude, ultimately recognising and conveying the idea that science is not "neutral" or "universal", but that it is a cultural phenomenon in itself. Experiments such as the one described in this volume by artist Joan Fontcuberta can raise awareness about the importance and cultural value of facts, precisely through the parallel presentation of the fictive alongside the "real".

The Exploratorium itself reflected Frank Oppenheimer's notion that people's perceptions and feelings were just as much part of the scientific process as were the physical aspects of the discipline. Cultural biases, social and economic pressure, the abilities and ideologies of individual scientists, the very structure of the scientific method have all shaped through time the way scientific research has proceeded and the way it was narrated and presented both inside and outside the scientific community. Yet there seems to be a misconception about scientific research: the idea that ethical issues only come to play in a final phase of the picture, when the scientific research has been carried out "neutrally" and there are decisions to be made about the "uses" and management of resulting technologies. In the real world, ethical components play a role at each and every stage of the research process, from the choice of the field of work to the shape and use of any resulting technological product. Artists often help uncover and describe such choices or directions of scientific research, not to shed a sceptical light upon the scientific process itself, but rather to reinforce facts by clarifying their source, background and standpoint.

Thus, alongside a specific work on the content of science museums, artists can provide an invaluable insight on the epistemology of science and work on bringing to light certain scientific-curatorial practices and narratives that tended to exclude many anthropological aspects of scientific research, collection and exhibition (Hermannstädter, 2019). An example of this self-reflection work is provided in this volume by Bergit Arends, who worked with artists at the Natural History Museum in London, with the aim of helping museum curators, collectors and researchers recognise present and past – often unconscious – non-inclusive practices.

A note on the structure of this volume

This work must be considered as a collection of stimuli that readers can access fluidly, following their personal interests and inclinations. Contributions have been grouped loosely into four sections, which should not be considered as rigid thematic separations. In fact, most papers include interconnected and cross-thematic reflections.

Each section of the book comprises both reflection chapters and case study charts.

Reflection chapters were written by curators, museum directors, artists, and educators. Their experiences and points of views reflect the plurality of perspectives surrounding the topic of art in science engagement intuitions.

Case studies were written by practitioners belonging to institutions from around the globe. The choice of projects is by no means exhaustive nor should it be considered as a collection of best practices. Instead, the selection aims to "photograph" a situation, exemplifying different cultural approaches and formats of art-science collaborations. Each described project is preceded by a short introduction by the editors, which alludes to a broader context or highlights a specific aspect of the described experience.

These short chapters are the result of synthetic transcriptions of charts in which authors were asked to give a very practical description of their projects, including reference to the involved partners and funders, the motivations and specific manner in which artists were involved, the practical outcomes and challenges of their experience.

The (open) conclusions invite readers to broaden the debate, whilst acknowledging the occurrence of curatorial practices that embrace what editors refer to as "Post-Disciplinary Approaches".

Notes

1 Today it is becoming increasingly difficult to make a distinction between science centres, technology museums, natural history museums, medical museums, maker spaces, university lab spaces, ecoparks, etc. The more generic expression "institutions dedicated to science engagement" helps us introduce the plethora of diverse institutions that come to represent the focus of this volume. A choice was made to include in the discourse "hybrid" institutions such as Science Gallery Dublin, Ars Electronica, FACT, SymbioticA, and others that have a clear value proposition

and explicit social mission connected to the engagement of different audiences with science, in a physical space, and through dedicated programming.

2 Many of the authors in this volume – including Hannah Redler Hawes, John Durant, Marina McDougall and James Peto – call for caution in the way art is brought into science engagement institutions. These experienced practitioners promote practices that avoid bringing together different areas of knowledge and competence superficially and try to establish deeper collaborations that foster a true dialogue between distinct epistemologies.

3 The acronym STEAM stands for Science, Technology, Engineering, Arts and Mathematics, which substitutes the traditionally used acronym STEM Science, Technology, Engineering, and Mathematics.

4 See Fox and Veeragoudar Harrell (2010) for NSF activity report and Glinkowski and Bamford (2009) for Wellcome Trust activity report.

5 This community created Report was compiled to advocate for the need of a full structured report and it contains a comprehensive list of prior third-party reports as well as 260 suggested actions for supporting transdisciplinary approaches.

6 Among the most known programmes are Arts at CERN; Art at ESO; NASA AIR programme.

7 Two examples of residency programmes for artist and scientists on "neutral grounds" are the Djerassi 'Scientific Delirium Madness' programme in the Santa Cruz Mountains and the "Project Dialogue" at the University of West England.

8 Weibel and Fruk (2013) have documented the recent advances of molecular art; Anker and Nelkin (2003) have investigated artists' fascination with modern biology; Daisy Ginsberg (Daisy Ginsberg *et al.*, 2014) has shown the collaborations between artists and designers in the realm of synthetic biology; Raimondi (2007) has gathered examples of artworks at the nanoscale.

9 For an interesting interview to Billy Klüver see Obrist (2003, pp. 509–523).

10 Reference is to the introductory panel to the artwork Lc.15:11–32 curated by the State Hermitage Museum.

11 In his chapter, James Peto gives an interesting account of how the integration of artists' works and perspectives has allowed to contextualise a rather "mute" piece of scientific equipment, changing radically the audience's perception of it.

12 "Gravity" was created by Tómas Saraceno in collaboration with the Italian National Institute for Nuclear Physics. It was shown at the MAXXI contemporary art museum in Rome (Italy) in 2018.

13 Extract from Conference programme: Art / Nature, *Contemporary Art in Natural History Museums and Collections* 26–27 June 2017. Available at: http://kunst.naturkundemuseum-berlin.de/wp-content/uploads/sites/3/2017/06/Reader_Art-Nature-Conference-2017.pdf (Accessed 22 July 2018).

Bibliography

Allen, S. & Gutwill, J. (2004) Designing with Multiple Interactives: Five Common Pitfalls, *Curator*, 47(2), 199–212.

Anker, S. & Nelkin, D (2003) *The Molecular Gaze: Art in the Genetic Age*, Cold Spring Harbor, NY: Spring Harbor Laboratory Press.

Byatt, A.B. (2000) Preface, in: Ede, S. (Ed.) *Strange and Charmed: Science and the Contemporary Visual Arts*, Lisbon: Calouste Gulbenkian Foundation.

Canadelli, E., Beretta, M. & Ronzon, L. (Eds.) (2019) Behind the Exhibit: Displaying Science and Technology at World's Fairs and Museums in the Twentieth Century, *The Smithsonian Institution Scholarly Press* (under publication).

Dabby, D. (2008) Creating Musical Variation, *Science*, 320(5872), 62–63.
Dewey, J. (1934) *Art as Experience*, New York: Berkeley Publishing.
Edwards, D. (2008) *ArtScience: Creativity in the Post Google Generation*, Cambridge, MA: Harvard University Press.
European Commission. (2012) *The New Renaissance: Will It Happen? Innovating Europe Out of the Crisis Third and Final Report of the European Research Area Board 2012*, Luxembourg: Publications Office of the European Union [Online]. Available at: https://ec.europa.eu/research/erab/pdf/3rd-erab-final-report_en.pdf (Accessed 22 July 2018).
Fox Harrell, D. & Veeragoudar Harrell, S. (2010) *Strategies for Arts + Science + Technology Research: Executive Report on a Joint Meeting of the National Science Foundation and the National Endowment for the Arts*, Washington, DC: National Science Foundation and the National Endowment for the Arts.
Friedman, A.J. (2010) The Evolution of Science Museums, *Physics Today*, 63(10), 45.
Gazzaniga, M.S., Ashbury, C. & Rich, B. (2008) *Learning, Arts, and the Brain: The Dana Consortium Report on Arts and Cognition*, New York: Dana Foundation.
Ginsberg, D. *et al.* (2014) *Synthetic Aesthetics: Investigating Synthetic Biology's Designs on Nature*, Cambridge, MA: The MIT Press.
Glinkowski, P. & Bamford, A. (2009) *Insight and Exchange: An Evaluation of the Wellcome Trust's Sciart Programme*, London: Wellcome Trust.
Hermannstädter, A. (Ed.) (2019) *Kunst/Natur: Interventionen im Museum für Naturkunde Berlin*, Berlin: Museum für Naturkunde.
Informal Science. (2017) *Art of Science Learning, Phase 2 – External Evaluation Final Report from a Summative Study of Impact* [Online], June. Available at: www.informalscience.org/art-science-learning-phase-2–external-evaluation-final-report-summative-study-impact (Accessed 10 July 2018).
Kahn, N. (2010) *Artist Statement* [Online]. Available at: http://nedkahn.com/statement.html (Accessed 01 August 2017).
Macdonald, S. & Basu, P. (Eds.) (2007) *Exhibition Experiments*, Hoboken, NJ: Blackwell Publishing Ltd.
Malina, R.F., Strohecker, C., LaFayette, C. & on behalf of SEAD network contributors. (2013) *Steps to an Ecology of Networked Knowledge and Innovation Enabling New Forms of Collaboration Among Sciences, Engineering, Arts, and Design, National Science Foundation* [Online]. Available at: https://seadnetwork.wordpress.com/report-and-meta-analysis/ (Accessed 03 July 2018).
McDougall, M., Bronwyn, B. & Semper, R. (2012) *Art as a Way of Knowing Conference Report*, San Francisco, CA: Exploratorium [Online]. Available at: www.exploratorium.edu/knowing/reports.html (Accessed 25 June 2018).
Merzagora, M. (2017) Science Centres and Science Engagement Activities as Research Facilities: Blurring the Frontiers Between Knowledge Production and Knowledge Sharing, *JCOM*, 16(2).
Merzagora, M. & Rodari, P. (2007) *La scienza in mostra: Musei, Science centre e comunicazione*, Milan: Bruno Mondadori.
Miller, A. (2015) *Colliding Worlds: How Cutting-Edge Science Is Redefining Contemporary Art*, New York: W. W. Norton & Co.
Morton, T. (2013) *Hyperobjects: Philosophy and Ecology after the End of the World*, Minneapolis: University of Minnesota Press.
National Economic Council and Office of Science and Technology Policy. (2015) *A Strategy for American Innovation*, Washington, DC: The White House [Online],

October. Available at: https://obamawhitehouse.archives.gov/sites/default/files/strategy_for_american_innovation_october_2015.pdf (Accessed 03 July 2018).

Obrist, H.U. (2003) *Interviews*, Vol. 1, Milan: Edizioni Charta.

Olmi, G. (1997) L'arca di Noè: La natura "in mostra" e le sue meraviglie, in: Basso Peressut, L. (Ed.) *Stanze della meraviglia: I musei della natura tra storia e progetto*, Bologna: CLUEB.

Oppenheimer, F. (1979) Aesthetics and the Right Answer, *The Humanist*, March–April.

Oppenheimer, F. (1972) The Exploratorium: A Playful Museum Combines Perception and Art in Science Education, *American Journal of Physics*, 40(7), 978.

Page, M. (2000) Creativity and the Making of Contemporary Art, in: Ede, S. (Ed.) *Strange and Charmed: Science and the Contemporary Visual Arts*, 104–119, London: Calouste Gulbenkian Foundation.

Raimondi, S. (2007) *Nanoart: Vedere l'invisibile/Nanoart: Seeing the Invisible*, Milan: Skira.

Root-Bernstein, R. & Root-Bernstein, M. (2001) *Sparks of Genius: The Thirteen Thinking Tools of the World's Most Creative People*, Boston, MA: Mariner Books.

Schirrmacher, A. (2013) Communicating Science: National Approaches in Twentieth-Century Europe, *Science in Context*, 26(3), 393–404.

Simons, N. (2013) Why Do We Interpret Art and Science So Differently? *Museum 2.0* [Online]. Available at: http://museumtwo.blogspot.com/2013/04/why-do-we-interpret-art-and-science-so.html (Accessed 10 July 2018).

Snow, C.P. (1959) *The Two Cultures and the Scientific Revolution*, Cambridge: Cambridge University Press.

Sullivan, G. (2005) *Art Practice as Research*, London: Sage.

Wagensberg, J. (2017) Museum of Art and Science, in: Grau, O. (Ed.) *Museum and Archive on the Move*, 25–36, Berlin: De Gruyter.

Wagensberg, J. (2005) The "Total" Museum, a Tool for Social Change, *História, Ciências, Saúde – Manguinhos*, 12(supplement), 309–321.

Weibel, P. & Fruk, L. (Eds.) (2013) *Molecular Aesthetics*, Cambridge, MA: MIT Press.

Weibel, P. & Latour, B. (2007) Experimenting with Representation: Iconoclash and Making Things Public, in: Macdonald, S. & Basu, P. (Eds.) *Exhibition Experiments*, 94–108, Hoboken, NJ: Blackwell Publishing Ltd.

Wilson, B., Hawkins, B. & Sim, S. (2013) A Transdisciplinary Approach to Art and Science Research: Permeable Research Frameworks, *The International Journal of the Arts in Society, Annual Review*, 7(1), 41–49.

Wilson, B., Hawkins, B. & Sim, S. (2012) Bayesian Provisionality Applied to an Epistemology of Science, *The International Journal of the Humanities: Annual Review*, 15–23, January.

Xanthoudaki, M. (2017) STEM to STEAM: Necessary Change or 'the Theory of Whatever'? *Spokes Magazine*, March 28.

Zilberg, J. (2015) A SEAD White Paper Working Group Meta-Analysis, in: Malina, R.F., Strohecker, C. & LaFayette, C. on behalf of SEAD network contributors (Eds.) *Steps to an Ecology of Networked Knowledge and Innovation Enabling New Forms of Collaboration Among Sciences, Engineering, Arts, and Design*, National Science Foundation [Online]. Available at: https://seadnetwork.wordpress.com/report-and-meta-analysis/ (Accessed 03 July 2018).

3 Premise

This first section includes two chapters that describe and interpret a number of historical facts and views which have paved the way to the present debate on art/science environments.

Pietro Greco describes early relationships between the fields of art and science, describing both the correlations and the existing divergences between the two worlds. He first shows broadly how art and science are products of biological as well as of cultural evolution. He then selects historical examples of ways in which art and science have mutually inspired one another in the Renaissance and later on in the 1900s. Greco concludes by recounting how, in history, the public perception of science has been profoundly shaped by artistic representations, which introduces the theme of art as an ally of science communication in its effort to democratise knowledge.

John Durant focuses more on the specific object of this publication, describing the cultural landscape of the 1960s, the concurrent birth of the science centre movement, and the more recent foundation of a number of transdisciplinary spaces. Opening with the case of the MIT museum in Boston (USA), he discusses examples of settings in which art and science were brought to interact at an "egalitarian" (non-ancillary) level. His synthetic and effective overview of recent history helps better understand the role and reciprocal influences between key personalities such as György Kepes, Frank Oppenheimer and Richard Gregory, at the origin of transdisciplinary Sciart. This chapter effectively introduces the current panorama, in which traditional institutions are questioning their role in society and new types of hybrid institutions are gradually emerging with the intent of enriching cultural conversations while reflecting and embracing a complex view of reality.

3.1 How art contributed to the public image of science

Pietro Greco

In the late 1950s and early 1960s, there was a lot of talk around a book by Charles Percy Snow (1959) denouncing a separation between "the two cultures" – scientific and humanistic. Snow strongly supported the idea that this tear had to be mended. And yet he took for granted that this separation had actually happened. Many intellectuals were shocked by Snow's provocation and reacted in their own way. In Italy, Primo Levi (1997) wrote that if there really did exist a "schism" between science and art, it was an "unnatural schism". He maintained that this separation was unknown to Dante and Galileo, just as it was unknown to Empedocles, Leonardo, Descartes, Goethe, Einstein, the anonymous builders of Gothic cathedrals, Michelangelo; to today's best artisans and to physicists, hesitating on the brink of the unknown.

Today, perhaps even more than in the past, we are forced to face Snow's provocation because the separation between the "two cultures" remains yet unhealed.

However, now more than ever before, it is useful to face this gap by adopting Primo Levi's approach. Because the "two cultures" are actually one. Levi's approach to the problem raised by Snow – the separation between the two cultures may exist, but it is unnatural – is extremely important because it tells us that art and science are different, yet deeply interpenetrated manifestations of a single culture: the human culture. Art and science intertwine and therefore influence each other in a much deeper way than many are led to believe. And the significant places of their interweaving and mutual influences are innumerable, impossible to summarise in a short chapter.

As a person concerned with science communication and with the construction of scientific imagery, I am interested in at least three dimensions of the dense, articulate and complex relationships between art and science: (1) Art and science as products of biological evolution; (2) Art and science as sources of mutual inspiration; and (3) Art as a significant allay in science communication.

Art and science as products of biological evolution

Solid research on the origin of art and of science in an evolutionary perspective allows us to identify a most likely common point of origin in the course of the evolution of the human species. In fact, art and science seem to be an expression of the new skills acquired by the species *Homo sapiens* – and probably by other species of the genus *Homo* (Hoffmann *et al.*, 2018). Specifically, of the ability to formulate abstract thinking and to use this skill to elaborate a complex representation of the world.

The graffiti left by *Homo sapiens* in the cave of Altamira, painted about 40,000 years ago, and those painted by Neanderthals in La Pasiega (Cantabria) and in Ardales (Andalusia) 65,000 years ago can be considered examples of the first manifestations of the artistic skills of two different species of the genus *Homo* and, at the same time, among the first (and convincing) manifestations of their ability to elaborate an abstract thought. *Neanderthals* and later *sapiens* reached levels of refinement unknown to other *Hominidae* and animals in their evolutionary (and therefore adaptive) ability to elaborate abstract models of the environment in which they lived some tens of thousands of years ago. An image of the world. And rock art is a plastic demonstration of these abilities.

This leads us back to another dimension of our research: the evolutionary origin of the aesthetic sense. A theme that goes well beyond mankind and its evolutionary history. It has in fact been demonstrated that other species, more or less phylogenetically close to man, possess an aesthetic intelligence and express it. For example, research on a chimpanzee called Congo seemed to reveal that our cousins can acquire a certain familiarity with abstract painting (Morris, 1967). When brushes and colours were offered to Congo, he used them to compose abstract figures that were somehow valued artistically by art critics. Moreover, Congo seemed to have a sense of the completeness of his work. When he finished painting he would put down his brushes and lose interest in the work he had produced. By contrast, when he felt he had not completed his artistic work, he continued undaunted and reacted badly if someone tried to take away his brushes and colours. Even birds, when they sing, seem to express a remarkable aesthetical sense. In an anecdotal story Mozart, on 27 May 1784 was said to have listened to the singing of a starling and then rushed home to transcribe the melody. On the basis of that musical line the great composer wrote the piano concerto number 17. Mozart had recognised in the chant of the starling an artistic sense. Something that was so significant from an aesthetic point of view that it deserved to be quoted and repeated (Frans de Waal, 2000).

On the other hand, does the gardener bird not show a notable aesthetic sense when building its nest? Mind you, an aesthetic sense that is perceived not only by humans, but also by the females of its own species, which choose male partners that build the most beautiful nests. The aesthetic sense of the gardener bird is not an anthropic reading of an ability that serves other biological purposes. The aesthetic sense of the gardener bird unequivocally emerges through real biological selection.

The issue of aesthetic sense and of its evolutionary meaning was already present in Charles Darwin's work that questioned, for example, the adaptive role of the peacock's tail (Darwin, 1871). Why does the male peacock have such a beautiful tail, which from a biological point of view seems like an impairment? A long and cumbersome tail certainly does not help in the struggle for survival. One would think that it was a feature destined to disappear through natural selection. Then why does the peacock have this magnificent and burdensome attribute? To answer this question, Darwin hypothesised the existence of another type of evolution, different from the natural selection of the fittest. He called it sexual selection. A process based precisely on the aesthetic sense. In this case on the aesthetic sense of female peacocks, that choose their partners on the basis of the beauty of their tail even if it is bulky and incongruous.

This is to say that the aesthetic sense is something that probably reaches far beyond the human species and belongs to a conspicuous part of the biological world. Understanding its origin and its function in biological evolution is important for humans, also to clarify the relationship between art and science. In fact, the aesthetic sense has played a profound role also in the evolution of science. It is no coincidence that the historian of physics Gerald Holton spoke of ionic seduction when talking about those men who – around the 5th century BC in Greece – began to use systematically the power of reason to investigate the universe and the cosmos, all harmoniously ordered (Holton, 1973).

Science, like art, is a product of cultural evolution that allows *Homo sapiens* to build an image of the world. So that another important dimension of the relationship between art and science in an evolutionary perspective concerns the psychology of research and, in particular, of creativity. In fact, there are (or seem to be) many common characters between scientists and artists. Many have attempted to study the mental mechanisms that lead scientists to investigate nature and formulate their findings. Among the first to address the problem of analogies (or homologies) between scientists and artists was a great French mathematician, Jacques Hadamard (1988), one of the fathers of the so-called theories of chaos.

Hadamard, in the early 1900s, identified two models through which the creativity of scientists acts: intuition and analysis. The intuitive model is substantially close to the creative model commonly adopted by artists. It feeds on analogies, metaphors, images, and mental experiments. The analytical model is the opposite, based on the rigid application of a formal logic, often of mathematical logic, which leaves (or seems to leave) little room for intuition.

But Jacques Hadamard convincingly demonstrated that even where there is a most rigorous adherence to the analytical model of scientific creativity there is an initial phase that involves intuition.

Albert Einstein – the greatest theoretical physicist of the 1900s and perhaps of all times – reflected on his personal research psychology. And he recognised that intuition was the trigger for his extraordinary creativity. The first step in the construction of his theories always consisted in imagining the physical reality he wanted to describe. And only after this first intuition would he create the necessary formal body describing a new theory. Einstein is a classic example of how the intuitive model of scientific creativity works (Miller, 1996).

In contrast, another great 20-century theoretical physicist, Paul Dirac, came to elaborate the quantum theory of the electron and discovered the physical reality of antimatter through a purely analytical procedure, by developing mathematical demonstrations. Dirac is a classic example of how the analytical model of scientific creativity works. And yet, one must remember that Dirac too was convinced of the rightness of his theories when he observed the simple and elegant "beauty" of his equations. Dirac, at the end of his theoretical work, granted something, by his own admission, to his aesthetic sense (Carafoli, 2013). And even before, if Hadamard is right, he must have owed something, perhaps without recognising it, to his intuition.

Intuition is, of course, the engine of artistic creativity. However, in the mental mechanisms that underlie the work of artists and scientists we can find something more than a marked analogy between science and art. Perhaps we can find the trace of their common origin: *Homo sapiens'* need and acquired ability to elaborate abstract thoughts and sophisticated representations of the world.

Art and science as sources of mutual inspiration

A second dimension in the relationship between art and science that interests those involved in communication and in the construction of individual and collective imaginaries is the fact that these two manifestations of culture represent a source of mutual inspiration.

It is quite simple to reveal how scientific concepts and scientific visions of the world have significantly influenced art. For example, the architecture historian Leonardo Benevolo (1991) has shown how the science of the 1500s and 1600s – with the discovery and affirmation of the new Copernican and Galilean universe (with the Earth revolving around the Sun and a practically infinite universe) – profoundly influenced the urban planning and architecture of the Baroque and of later eras.

Art re-proposed the new scientific image of the world elaborated by Galileo Galilei (Greco, 2013a). The gardens of Versailles are an example of the way in which architects expressed the "capturing of the infinite", reinforcing this concept among the crowds.

The profound influence of scientific discoveries and, to use Galileo's words, of "scientific speculations" can also be found in literature. Italian writer and literary critic Italo Calvino (1967) claimed that Italian literature from Dante to Galileo was marked by a "profound vocation": natural philosophy. He saw literature as a map of the world and knowledge.

Similar attitudes can be found in all European literature, but it seems to be particularly evident in Italian literature until well after the 1800s, influencing Leopardi and Calvino himself. In fact, Calvino maintains that when Italian literature loses its peculiar vocation it loses its identity and strength.

Yet it is not only science that influences art. Art also influences science. We have already mentioned the aesthetic criterion used by Paul Dirac to establish the validity of his scientific theory in the absence of experimental proof. But there are also much more direct inspirations. The biology historian and evolutionist theoretician Stephen J. Gould (Gould & Lewontin, 1979) was explicitly inspired by the Basilica of San Marco in Venice and by the internal structure of its dome. The architecture helped him elaborate the concept of "contingency" and of its role in the structure of biological evolution.

Science returned to Europe at the beginning of the 1400s through Florentine "artist" workshops, those of Filippo Brunelleschi, Masaccio, Donatello. And also through the studies of Piero della Francesca, Leonardo da Vinci, Michelangelo Buonarroti, Raffaello Sanzio. The artistic Renaissance was also and above all a scientific Renaissance. We can say, with good reason, that Europe discovered the scientific way of constructing images of the world thanks to art.

Very often art and science reflect, almost in parallel, the "spirit of the times". And they reinforce it. This synchronisation is sometimes astonishing in its perfection.

Let us consider the "strange case" of Einstein and Picasso. In 1905 Albert Einstein was a 26-year-old young man with a degree in physics and an employee of the Patent Office in Bern, a stranger to the academic world. Yet he managed to elaborate a physics theory, the so-called theory of special relativity, with which he swept away the concept of absolute space and time in science.

A few months later, in 1906 in Paris, another man of the same age, Pablo Picasso, began to work on *Les Demoiselles d'Avignon* (completed in 1907), This painting swept away the concept of absolute space from figurative arts. The painting is probably the result of a thought that developed quickly. In fact, at the beginning of 1906, Picasso had painted *Harem*, where the same subject – a group of women in a closed house – was represented with a classical perspective, based on the notion of absolute space. After a few months those women were placed in the relativistic space of *Les Demoiselles d'Avignon*. Something must have happened.

A critique of the concept of absolute space was actually maturing in the European intellectual landscape. However, the coincidence of two young men, Albert and Pablo, having similar intuitions in two such different cultural dimensions is truly extraordinary. Also and above all because Einstein and Picasso did not know each other nor their respective work. There must have been something that influenced them: and this something was the ability to grasp "the spirit of the times" and to bring it to a synthesis. In agreement with others, the English historian Arthur Miller (2001) argued that this common perception was mediated by the work of French mathematician Henri Poincaré, that was known to both Einstein (directly) and Picasso (indirectly).

Art and the public image of science

Finally, there is a third dimension which is worth considering: that of art as a significant medium for science communication. In this perspective, the first aspect concerns the fundamental role assumed by artistic culture in the construction of the collective and individual scientific imaginary.

For example, according to Katherine Hayles (1991), the relationship between science and literature happens at three levels, which relate to (a) rhetoric – with mutual exchange of communication registers; (b) concepts – with an exchange of themes, metaphors and analogies; and (c) a profound culture. There seems to be what Eugenio Montale defined as an obscure pilgrimage of fertile ideas and epistemological instruments that move incessantly from one field to the other and which shape the cultural matrix in which we live. Gerald Holton (1973), physics professor and science historian, calls *themata* the objects of this obscure pilgrimage. He argues that the exchange between the epistemological tools of science and literature contributes to that complex and radical metaphorical reorientation which brings – in science, but more generally in culture – to a "paradigm shift".

As an example, one could consider the case of the *Sidereus nuncius*, written by Galileo Galilei in 1610, after observing through a telescope the rough surface of the Moon, the satellites surrounding Jupiter, the infinite stars. At the beginning of the 1600s this scientific book becomes the central object of one of the first documented mass culture phenomena. Historian Elizabeth L. Eisenstein (2005) noted that the *Sidereus Nuncius* was a sensational event. Not only did it turn its author into an international celebrity, but it also did for astronomy what Luther's first books had done for theology: it provoked literary excitement and generated a new type of publicity. In short, it contributed more than any other book of its time to giving astronomy (and, consequently, the new science) a mass-culture dimension. And why did this happen? Because the book was in line with the spirit of the times. It was in tune with the baroque hunger for novelty and wonder at all costs: 'Discover new worlds or drown!' wrote the poet Antonello Chiabrera (Greco, 2013a).

Galileo was indeed talking about incredible novelties: he depicted the ultralunar world as the terrestrial world: corruptible and imperfect. He thus recomposed the physical unity of the universe and showed that there was at least one other planet, Jupiter, around which moons rotated. This meant that, by analogy, it was possible that the Copernican model of the universe was not a mathematical artifice but a physical reality. He demonstrated that there are many more fixed stars than those visible to the naked eye, and therefore that the universe is not closed and finite as it appears in the Aristotelian-Ptolemaic description, but indeed huge and perhaps infinite.

All these novelties became part of a mass culture thanks to the immediate action of poets, writers, and painters. Ludovico Cardi da Cigoli, a friend of Galileo, was certainly inspired by the drawings of the Florentine physicist when he depicted the moon (with spots, craters and bumps) in the Assumption of the Virgin frescoed on the dome of the Pauline Chapel of the Church of Santa Maria Maggiore in Rome.

Andrea Battistini (1989) describes how Adam Elsheimer depicted the new "Galilean sky" in his most cherished work "Fugue in Egypt", where the sky was painted with a naturalism that was unknown to the world of art at the time, with the eye of an experimenter looking at the universe through a telescope.

Galileo himself became a reference point for a large group of poets and, together with his work, was placed at the centre of a series of poems (some encomiastic) such as those by Giambattista Manso, Giovanni Battista Marino, Gabriello Chiabrera, Johannes Faber of Bamberg, Giuseppe Battista and John Milton in his Lost Paradise.

'*Vicisti, Galileae*!' wrote Scottish poet Thomas Seggett in a poem in Latin that was later quoted in the *Narratio de Jovis Satellitibus*, written in September 1610 by Kepler (1611) in honour of Galileo. It should be noted that less than seven months had passed since the publication of Galileo's book and already the *Nuncius* had rebounded several times between the world of science and art and not everyone was an ecstatic singer of the Galilean announcement. An excerpt of a tormented poem written by John Donne (1997) in 1611 reads:

> 'And new philosophy calls all in doubt,
> The element of fire is quite put out,
> The sun is lost, and th'earth, and no man's wit
> Can well direct him where to look for it'.

But, beyond the nature of the emotions it provoked, it is evident that the *Sidereus Nuncius* affected not only experts, "philosophers and astronomers", but the entire European mass culture. Even numerologists were inspired by the discovery of the four Medicean planets and weaved a new networks of numerical correspondences around the cosmos and the essence of man.

The dissemination of the message contained in the *Sidereus Nuncius* was re-launched, interpreted, amplified through "other means" by the most diverse groups of intellectuals in the baroque era. And it was precisely this action, on closer inspection, that turned Galileo's observations into "culture", quickly contributing to the shaping (or rather, to the re-shaping) of the collective scientific imaginary while the new science was still forming.

Galileo was well aware of the importance, for science, of this process and he wrote to Thomas Stigliani that the poet was also to cultivate scientific speculations ("Si concede anco al Poeta il seminare alcune scientifiche speculazioni") (Greco, 2013a).

An example of such "scientific speculations" is the front page of the *Telluris theoria sacra*, where Thomas Burnet (1681) wants to convey the idea that to develop a theory of terrestrial evolution one needs both the metaphor of linear time and that of circular time. Once again, the artistic image becomes a powerful vehicle for transmitting "scientific speculations".

Even in modern times the use of artistic metaphors is quite common. For example, in Italy, Oliviero Toscani's advertising campaign with photographs of an AIDS patient has greatly contributed to the remodelling of the imagery related to this infectious disease.

Another example is how the imagination surrounding the condition of "madness", has been continuously stimulated, throughout history, by its artistic representation. In the Ship of Fools, painted by Hieronymus Bosch between 1490 and 1500, there is a clear representation of the separation of fools from society. A separation that never actually occurred in peasant society.

Oppositely, in more recent times, photography was a powerful tool for denouncing the exclusion of "lunatics" from the rest of society. In photos by Luciano d'Alessandro (Bersani, 2015), for example, we can witness the exclusion and isolation of madmen in mental hospitals. After the Basaglia law in Italy, in 1978, communication barriers dropped and photography reinforced the idea that it is impossible to clearly distinguish between madness and "normality".

Conclusions

If one takes into account how art contributes constantly to the (re)building of our scientific imagery, one must consider its place in the public communication of science. A communication that has taken on a decisive role today both for the development of scientific knowledge and for the democratic development of society.

In fact, this is a period of transition in the way scientists work. From the area of academia, we are now moving into a post-academic era of science. The change is leading to the revision or reinterpretation of some founding values of scientific activity.

One of the main characters of this transition is that in the academic era the decisions relevant to the development of science were taken almost exclusively within the scientific community. In those that, not surprisingly, were called the "ivory towers".

In the post-academic era, decisions relevant to the development of scientific knowledge must be shared between scientists and a number of non-expert audiences (Ziman, 2000).

We have therefore moved from a world in which science and society were essentially separate, to a world in which science and society are strongly interpenetrated. In this context, the issue of public science communication changes strongly. If once communication was mainly internal and took place between scientists, today the communication of science involving non-expert audiences has also become deeply relevant.

The problem affects us all closely. Because in this same post-academic period science has increasingly become one of the fundamental levers of social dynamics: it is the source of knowledge on which technological innovation systematically relies (Greco, 2013b). Today we see that scientific issues involve politics and society across the board. In the United States an aggregating factor behind new political majorities are a precise bioethical vision and some precise (anti)scientific visions: from that of creationism, opposing the Darwinian theory of biological evolution, to that of climate change denial.

We are all, in one way or another, summoned to make important decisions on the most varied scientific topics: from stem cells and cloning, to changing climate and the control of new weapons of mass destruction. Topics with a scientific value are, more and more often, at the top of political and social agendas. On the other hand, it is difficult to interpret the contemporary world without having the slightest ability to interpret the evolution of science (and of the technology that is based on science). Consequently, communication around these subjects has become a major problem not only for scientists, but for society as a whole.

In short, today, a fully democratic society is a society that has great awareness of the scientific issues and of the communication processes that affect them.

Art is one of the main tools chosen by man to interpret reality and to communicate. For all these reasons and others, in the post-academic era of science, art becomes one of the main channels through which "scientific speculations" are disseminated. It is in this frame that the relationship between science and art, which has always been important for the cultural development of society, has become an important element for the development of democracy itself.

Bibliography

Battistini, A. (1989) *Galilei*, Roma: Laterza.

Benevolo, E. (1991) *La cattura dell'infinito*, Roma: Laterza.

Bersani, A. (2015) Fotografia e Follia, *MemeCult* [Online]. Available at: www.memecult.it/fotografia-e-follia/ (Accessed 25 June 2018).

Burnet, T. (1681) *Theoria Sacra*, London: Kettilby.

Calvino, I. (1967) Philosophy and Literature, *The Times Literary Supplement*, 28 September, 871–872.

Carafoli, E. (2013) Bellezza e verità: Arte e scienza a confronto, in: Greco, P. (Ed.) *ArmonicaMente*, Udine: Mimesis.

Darwin, C. (1871) *The Descent of Man, and Selection in Relation to Sex*, London: John Murray.

De Waal, F. (2000) *The Ape and the Sushi Master Reflections of a Primatologist*, New York: Basic Books.

Donne, J. (1997) An Anatomy of the World: The First Anniversary: To the Praise of the Dead and the Anatomy, in: Smith, A.J. (Ed.) *The Complete English Poems*, Harmondsworth: Penguin.

Eisenstein, E. (2005) *The Printing Revolution in Early Modern Europe*, New York: Cambridge University Press New York.

Gould, S.J. & Lewontin, R.C. (1979) The Spandrels of San Marco and the Panglossian Paradigm: A Critique of the Adaptationist Programme, *Proceedings of The Royal Society*, 205, 581–598.

Greco, P. (2013a) *Galileo, l'artista toscano*, Milano: Springer.

Greco, P. (2013b) Prefazione, in: Bush, V. (Ed.) *Manifesto per la rinascita di una nazione*, Torino: Bollati Boringhieri.

Hadamard, J. (1988) *Thematic Origins of Scientific Thought: Kepler to Einstein*, Cambridge, MA: Harvard University Press.

Hayles, K. (1991) Literature and Science, in: Coyle, M. *et al.* (Eds.) *Encyclopedia of Literature and Criticism*, London: Routledge.

Hoffmann, D.L. *et al.* (2018) U-Th Dating of Carbonate Crusts Reveals Neandertal Origin of Iberian Cave Art, *Science*, 359, 912–915.

Holton, G. (1973) *Thematic Origins of Scientific Thought: Kepler to Einstein*, Cambridge, MA: Harvard University Press.

Kepler, J. (1611) *Dissertatio cum nuncio sidereo: Narratio de observatis Jovis satellitibus*, Frankfurt: Sumptibus Z. Palthenii.

Levi, P. (1997) *Opere*, Torino: Einaudi.

Miller, A.I. (2001) *Einstein, Picasso: Space, Time, and the Beauty That Causes Havoc*, New York: Basic Books.

Miller, A.I. (1996) *Insights of Genius: Imagery and Creativity in Science and Art*, New York: Copernicus.

Morris, D. (1967) *The Naked Ape: A Zoologist's Study of the Human Animal*, New York: McGraw-Hill.

Snow, C.P. (1959) *The Two Cultures and the Scientific Revolution*, Cambridge: Cambridge University Press.

Ziman, J. (2000) *The Real Science: What It Is and What It Means*, Cambridge, UK and New York: Cambridge University Press.

3.2 Making meaning with art, science and technology

John Durant

A Museum of . . . what, exactly?

People who have never visited the MIT Museum often ask me: 'What is your museum about?' I am slightly embarrassed to admit that I always find this question hard to answer. For although our museum is not large, at any given time it tends to offer exhibitions on what appear at first sight to be an assortment of wildly different subjects. Among the offerings of 2017, for example, are the following: *Images of Discovery*, an exploration of the role of image making in research; *Cosmic Bell: Exploring Quantum Weirdness*; *The Enemy*, a virtual reality experience of human conflict based on the work of photojournalist Karim Ben Khelifa; and longer-lasting galleries devoted to holography, kinetic sculpture and robotics/artificial intelligence.[1]

Our collections, too, are extraordinarily diverse. They include architectural and design drawings, cathode-ray tubes, computers, engineering design models, gene sequencers, gravity wave detectors, holograms, musical instruments, paintings, photographs, scientific toys and games, sculptures, slide rules, space suits and yacht models.[2] The Museum's Collections Committee frequently approves the acquisition of artefacts from across the entire spectrum of STEAM – Science, Technology, Engineering, Art and Mathematics. Indeed, some acquisitions – a recent case in point is the David A. Huffman Collection of paper flat folded models, sketches, notes and calculations – are best understood as STEAM objects in their own right.

The reason for all of this diversity is simply stated. The MIT Museum is not a museum of art, nor a museum of science, nor yet a museum of technology. Rather, it is a museum of a great research institute that has always taken a catholic view of the nature of its proper pursuits. The Massachusetts Institute of Technology's founding document, William Barton Rogers' *Objects and Plan of an Institute of Technology* (1860), refers to 'A Society of Arts, a Museum of Arts, and a School of Industrial Science'. These terms evoke a mid-19th-century culture in which the boundaries between art, science, and technology, were nothing like so clearly drawn as they are today; and, true to this originating vision, MIT has always interpreted the 'T' in its title very broadly, to include the arts and the sciences, and also (in the 20th century) the humanities and the social sciences.

A case in point is MIT's Center for Advanced Visual Studies (CAVS), whose 50th anniversary was celebrated in 2017. CAVS was the brainchild of the Hungarian born artist György Kepes. After joining the MIT faculty in 1946, Kepes became steadily more interested in the dialogue between art and science; and upon its founding in 1967, CAVS quickly established a reputation as a place where artists who were interested in science and scientists who were interested in art were assured of a warm welcome and encouraged to develop collaborative projects across disciplines. Early among those who were attracted to CAVS as Center Fellows was the co-founder of the influential Group ZERO, German artist Otto Piene; and under the leadership of founding director Kepes (1967–1974) and Piene (1974–1994), CAVS established a distinctive and enviable reputation for art in dialogue with science and technology.

Some of the fruits of this dialogue are on display in the museum, in the form of art works by CAVS artists Nam June Paik and Charlotte Moorman (*TV Cello*, 1971), Wen-Ying Tsai (*Large Tuning Fork*, 1972), Christopher Janney (*Soundstair*, original artwork 1978), Harriet Casdin-Silver (*A Woman III*, 1979), Lowry Burgess (*The Boundless Cubic Lunar Aperture*, 1989) and others. Most conspicuous of all, and immediately evident on entering the museum, is the 'starburst' of Piene's *Cereus* (2008), a gigantic inflatable sculpture made from parachute nylon. 'Why is there no art in space, why do we have no exhibitions in the sky?' Piene had asked; and in 1968 he answered his own question by introducing "Sky Art" to the world. Forty years later, in 2009, Piene oversaw the creation of *Cereus* and *Paris*, based on his original drawings. They were featured in the major public arts event of MIT's 150th anniversary celebration in spring 2011, when thousands gathered in the Institute's Killian Court before the Great Dome to help inflate and guide these beautiful rising stars high into the night sky.

Re-thinking the art-science-technology relationship

The experience of leading the MIT Museum has steadily changed the way I think about the relationship between art, science, and technology. Coming here from two previous positions in museums that saw themselves as specialising in science and technology, I was accustomed to seeing art as something ancillary rather than integral to the work at hand. To be sure, many science and technology museums have art works in their collections; but in the main, these objects have been acquired largely for their relevance to core curatorial and interpretive interests that lie elsewhere. Thus, paintings and posters are to be found in the collections of railway museums, where they illustrate topics in the history of transportation; and botanical paintings are commonly collected by natural history museums, where they illuminate the history of natural history. Even where art objects come to have iconic status in such museums – a case in point is Harvard University's famous "glass flowers" collection – they are generally found to be there in service of other (however obscure) scientific and pedagogical interests.[3]

This view of art as ancillary to science and technology is totally insufficient. Of course, art *can* be made to support scientific and technological storytelling – or any other kind of storytelling, for that matter. (Obviously, a large part of the history of western art is a history of art in service to the Church, the State, and/or the whim of wealthy patrons.) Sometimes, great works are produced in the course of such artistic servitude; but all too often, the results are formulaic and derivative. Not least, this is because putting art in service of something else inevitably constrains and limits it. Given its own space and freedom to move, art is a field of enquiry in its own right. What this exactly means – that is to say, defining more precisely the field that art takes for its inquiries, and describing as clearly as possible the various methods artists employ in going about their work – has been the subject of almost endless art-historical and critical debate; and (with just one exception, of which more later) I will not delve into that literature here. Suffice it to say that art is best understood in its own terms as a creative meaning-making activity, and not in utilitarian terms, as an undertaking to be evaluated primarily according to how well it contributes to some other endeavour.

Having freed art to be itself, however, it is important to recognise that artists claim the right to respond to anything and everything around them, including not only the objects of science (the found world) and technology (the made world), but also science and technology themselves, where these are understood as human and cultural endeavours. Art that is not subordinated to the interests of science and technology is well placed to engage creatively with these things, generating meaningful interventions that perhaps neither artists nor scientists nor engineers working alone would likely have produced. Art-science-technology collaborations of this kind are a very distinctive feature of the cultural landscape of the past half century or so; and this, more than anything else, accounts for the place of art in the collections and displays of the MIT Museum.

The case of György Kepes is instructive. He was originally recruited to MIT to teach visual design to architecture students. Profoundly influenced by the Bauhaus artist László Moholy-Nagy, with whom he had shared a commercial design practice in Berlin, Kepes was deeply interested in visual communication, which he saw as being a universal – and therefore, by implication, international – language.[4] At MIT, Kepes found common cause with many scientists and engineers, and in 1956 he published *The New Landscape in Art and Science*, which had its genesis in an exhibition he had organised for Harvard's Carpenter Center five years earlier. The book paired art images from all periods with scientific images made with the help of a variety of contemporary imaging technologies, including radar, stroboscopes and x-ray machines (Kepes, 1956). *The New Landscape* opened a space – subsequently occupied by many others – in which scientific images are considered as ‘art objects’ in their own right; and it paved the way for various types of art-science-technology collaboration that became the hallmark of CAVS more than a decade later.

Kepes’ motives in engaging with scientists and engineers are discernible from his published remarks, at the time and afterwards. In the Dedication Catalogue of CAVS (1967), he spoke about the need for ‘an art of public forms and corresponding dimensions’, and again of art, ‘fast approaching the environmental scale and by its own inner dynamics as a craft becoming a collaborative enterprise involving science and engineering’. Underlying this vision was a deep discontent with what Kepes often referred to as the “chaos” of a technological society in which relationships both among people and between people and their larger (natural, and constructed) environment had become deeply disturbed. In 1972, CAVS artists became involved in a large-scale civic art project around attempts to clean up the heavily polluted Charles River that flows between Boston and Cambridge, Massachusetts. In an interview recorded at the time, Kepes described the motives for his Center’s involvement in this project as follows:

> There is an obvious and sometimes painful discrepancy between life as it is and life as it should and could be. But most of us retain still a confidence in life, and try to find ways to bridge this gap. Artists are among those who have a passionate commitment to the completeness of life. Today one doesn’t need to have too much imagination or extra sensitivity to see how our urban life we created for ourselves is not living up to its potentials.
>
> (bobvidpix, 2011)

Kepes had multiple reasons for engaging with scientists and engineers. For one thing, science and technology offered new means (materials, media, techniques, technologies) for the accomplishment of large-scale art works; and for another, they offered new motives for such works, through the multiple disruptions and dislocations that Kepes perceived to be the result of myriad large-scale applications of science and technology in the wider world. In some ways, Kepes’ art-science-technology collaborations of the late-1960s and early 1970s are best understood as invitations to scientists and engineers to join in the effort of imagining better, more fully human ways of implementing their own expertise in the world – better, that is, than many of the ways that were being attempted all around them without the help of the arts.

Art and science at the birth of science centers

At around the time that Kepes was creating CAVS, the first stirrings of what would become the international science center movement were taking place; and here, too, we can discern moments of convergence between art and science. A prime example is to be found in the work of Frank Oppenheimer and his collaborators in creating the Exploratorium in San Francisco (founded, 1969). On this occasion, of course, it was not an artist who was reaching out to scientists and engineers, but rather a scientist and engineer who was reaching out to artists. As Oppenheimer's biographer K. C. Cole makes clear, however, Oppenheimer's love of art was every bit as deep as his love for science; and it was, to a great degree, inherited: his mother Ella Friedman was a painter; and his father, Julius, was a collector who filled the family home in New York with art works and helped to set up that city's Museum of Modern Art (Cole, 2009, p. 27).

Familial influence helps to explain why art was always integral to Oppenheimer's vision for a new kind of museum; but another, more fundamental part of the explanation has to do with Oppenheimer's conception of what his Exploratorium was all about. Here is Cole, describing her first visit to the place in San Francisco's Palace of Fine Arts in the early 1970s:

> It took a while for the eyes to adjust – and not only in a literal sense. The strange hums and flickers gave the place an ethereal, otherworldly feel. There was a big yellow tent on my left that housed an electronics shop. On my right was a noisy machine shop, all sparks and whines of saws and welding and lathes. Hanging above the shop, a handmade sign announced: *here is being created the exploratorium, a community museum dedicated to awareness.*
>
> (Cole, 2009, pp. 4–5)

At face value, this seems a surprising way for the man who is generally recognised as the founder of the international science center movement to define his mission. And yet it was true that Oppenheimer always saw the Exploratorium as being about something greater than science itself. Above all, it was intended to help visitors think curiously and critically about their experiences of the world; to help them question "common sense" and received wisdom wherever these things stood in the way of a deeper grasp of underlying realities. In this, Oppenheimer was almost certainly influenced by his own experience as a young physicist and (alleged) Communist, who had himself been a victim of McCarthyist political persecution. As the physicist Murray Gell-Mann has noted, the ultimate goal of the Exploratorium 'was to get people so addicted to understanding that they would somehow become inoculated against the clever deceptions of some advertisers and politicians' (Murray, 2009, p. xiv).

Oppenheimer's focus on perception and his enthusiasm for combining insights from art and science gave him much in common with the English neuropsychologist Richard Gregory. Gregory had served with the Royal Air Force's Signals branch during World War II. But at the end of the war, he won a place at Cambridge University to read experimental psychology, and there he developed a special interest in visual perception. In a series of early studies, he argued (following the 19th-century German physiologist Hermann Van Helmholtz) that the process of seeing was active, rather than passive. In 1973, Gregory collaborated with the distinguished art historian Ernst Gombrich on an exhibition at the Institute of Contemporary Arts (London)[5] and an accompanying book called *Illusion in Nature and Art* (Gregory & Gombrich, 1973).

In his earlier much celebrated work, *Art and Illusion* (1960), Gombrich had presented a constructive view of perception to account for the fact that artists in different times and places have represented the visible world so differently (Gombrich, 1960). In Gombrich's view, the artist brings prior mental "schemata" to his or her engagement with a visual subject, and then works to adjust these schemata to accord with what is seen. It is a fair summary of Gregory's scientific work on visual perception to say that it was a systematic exploration of the cognitive psychological foundations of this sort of constructive process, now understood not as a special feature of visual art but as a general characteristic of the human visual system.

Gregory and Oppenheimer were kindred spirits. Both were polymaths, both displayed an infectious curiosity about the world, and both were widely considered (by their admirers, anyway) as delightful eccentrics. Gregory brought to San Francisco his developing fascination with visual illusions of all sorts, and according to Cole he had an enormous influence on both the early perception displays at the Exploratorium and Oppenheimer himself (Cole, 2009, p. 176). Gregory, too, learnt much from Oppenheimer – so much, in fact, that he went back home and created his very own place, which he named – in open acknowledgement of Oppenheimer's precedent – the Bristol Exploratory (founded, 1987).[6]

Like Oppenheimer, Gregory used the Bristol Exploratory as an experimental site for creative work at the interface between art, science and technology; and among other things, he showcased there his ongoing work on visual illusions.

Second only to Oppenheimer himself, Gregory was enormously influential in the birth of the international science center movement. Like his American counterpart, Gregory was simultaneously a practitioner of, an advocate for, and a theorist about a new way of engaging visitors with science and technology. He rooted his entire approach to interactive exhibits in a philosophy of "hands on science" (the term is Gregory's) that was firmly rooted in his distinctive cognitive psychology.[7] Since Gregory's cognitive psychology itself was developed under the influence of a distinguished art historian, it seems entirely reasonable to claim that the international science center movement arose in part out of a post-war dialogue between art (or at any rate, art history) and science.

From art-science-technology to SciArt

The field of art-science-technology collaboration received a big boost in 1996, when the Wellcome Trust in London launched its so-called *SCIART* funding program for creative partnerships between artists and scientists.

These partnerships were aimed at producing work, 'that simultaneously promotes the public appreciation of medical science and encourages innovation in the visual arts' (Wellcome Trust, 2018). This was the start of a ten-year program of funded work that did much to establish the emergent field of SciArt nationally in the UK, and to a degree internationally. In their recent review of this program, Charlotte Sleigh and Sarah Craske suggest that it began by funding generally little-known artists, and that over the course of the decade the art produced tended to be subordinated to the interests of science communication. Thereafter, however, they trace a trend to consider SciArt (and what they term its newer, more rebellious sibling, "BioArt") as a "transdiscipline" preoccupied above all with the solution of so-called wicked (highly complex, mixed natural/social) problems (Sleigh & Craske, 2017).

Transdisciplinary SciArt is a distinctive feature of a number of new galleries that have opened over the past decade. Two examples here must suffice. First, in 2008 Trinity College Dublin opened its Science Gallery. From the outset, Science Gallery founder Michael John Gorman set his sights on creating "a meeting place for ideas". Unencumbered by either collections (read, science museums) or the need to attract children (read, science centres), he decided that the Science Gallery should be located 'at the membrane between the College and the Community'. It would be built around a good café, it would be targeted at young adults, and it would aim 'to surprise, to provoke, and to seduce people to deepen their participation'.[8]

Science Gallery Dublin opened in February 2008 with an exhibition called *Lightwave: Ready to be Illuminated?* Since then, it has kept up a steady drumbeat of three to five special exhibitions per annum, each the dream child of a transdisciplinary *Leonardo Group* of scientists, engineers, artists and designers, and each featuring a rich array of internationally crowd sourced SciArt exhibits.[9] So great was the initial impact of this new approach that in 2012 a Global Science Gallery Network was launched, with the aim of establishing Science Galleries based on the Dublin model in eight cities around the world by 2020. At the time of writing, the Global Science Gallery Network has six members – in Dublin, London, Melbourne, Bengaluru, Venice and Detroit. Time alone will tell how far and wide this approach to art, science and technology will spread.

My second example of SciArt in unconventional galleries is an initiative of the Wellcome Trust. Following an earlier and relatively unsuccessful attempt to create a gallery about medical science for the public in its headquarters building in London's Euston Road, in 2007 the Trust launched Wellcome Collection, a museum and library aimed principally at an adult audience and billed as 'the free destination for the incurably curious'. At the heart of the great success of Wellcome Collection (it is currently attracting around 700,000 visitors a year) has been a series of arresting special exhibitions, many of which have brought together the work of artists, historians and scientists to address edgy themes in a typically transdisciplinary way.

A good example of the Wellcome Collection's distinctive approach to SciArt is provided by its recent exhibition: *Making Nature: How We See*

Animals, which ran from 1 December 2016 through 21 May 2017. In the words of the Wellcome Collection, *Making Nature* examined, 'what we think, feel and value about other species and the consequences this has for the world around us' (Wellcome Collection, n.d.). To this end, the exhibition brought together an astonishing mix of artefacts from literature, film, taxidermy and photography (among other media), in ways that generally defy any neat classification under headings such as "art", "science", "history", or "natural history".

In the first main section of *Making Nature*, called "Ordering", the most striking feature is a video installation by the Puerto Rico–based artists Jennifer Allora and Guillermo Calzadilla, in collaboration with the American writer Ted Chiang. *The Great Silence* is set in the Arecibo Observatory in Puerto Rico and the forests immediately around it, which are home to the endangered parrot, *Amazona vittata*. The genius of the work lies in its noticing the ironic conjunction of two apparently unrelated things: a radio telescope that is used in the ongoing scientific search for extra-terrestrial intelligence; and a nearby highly talkative parrot that is threatened with imminent extinction. This conjunction is captured on film, and through an accompanying narrative (composed by Chiang) that is told in the first person by the parrots themselves. This narrative starts as follows:

> The humans use Arecibo to look for extraterrestrial intelligence. Their desire to make a connection is so strong that they've created an ear capable of hearing across the universe.
>
> But I and my fellow parrots are right here. Why aren't they interested in listening to our voices?
>
> We're a non-human species capable of communicating with them. Aren't we exactly what humans are looking for?
>
> (Allora & Calzadilla & Chiang, 2015)

The title of *The Great Silence* is a reference to the so-called Fermi paradox, named after physicist Enrico Fermi, who pointed to the apparent contradiction between the high probability generally attached by scientists to the existence of extraterrestrial civilisations, and the fact that to date there is absolutely no evidence for such civilisations. 'Where is Everybody?' Fermi famously asked an informal group of colleagues in 1951.[10] To this question, Allora and Calzadilla and Chiang's parrots' laconic answer is clear:

> We're here, right under your noses . . . but not for much longer.

Coda

I like to think that Kepes and Oppenheimer and Gregory would have enjoyed *The Great Silence*. This work brings artistic sensibilities to some of the great scientific and social questions of the 21st century, in ways that are designed both to disturb and to provoke. Art that is genuinely engaged with science and technology has much to offer, precisely because it imposes new candidate meanings upon these subject areas of which we, the citizens of highly scientific and technological societies, are badly in need. The contributions of art are not always (though, as we have seen, they may be) morally challenging. At other times, they may evoke surprise, bemusement, amusement, astonishment, or wonder. Whatever its specific contributions may be, however, art that is genuinely engaged with science and technology always adds to the richness of the conversations that we have and can have about the worlds of our discovery and our making.

Acknowledgments

I would like to thank Dr Michael John Gorman, the founder director of Science Gallery, Dublin, for agreeing to be interviewed, and my colleagues Seth Riskin and Gary Van Zante for their comments on an earlier draft of this chapter.

Notes

1 See details of these and other recent MIT Museum exhibitions (MIT Museum, 2019a).
2 See details of the MIT Museum's collections (MIT Museum, 2019b).
3 For an excellent account of the curious origin and enigmatic status of these objects, see Lorraine Daston, *The Glass Flowers* (Daston, 2004).
4 György Kepes published a series of influential works on this subject, starting with *Language of Vision* (Kepes, 1944).
5 The exhibition travelled widely in Britain and North America through the middle years of the 1970s.
6 The Bristol Exploratory was the first independent science center in the UK, and one of the first in Europe. It closed (for financial reasons) in 1999, and was succeeded by Explore@Bristol, which continues Gregory's legacy of hands-on science in Bristol to this day.
7 Gregory claimed to first see the importance of tactile experience to visual understanding through his early case study of an adult who recovered from early blindness. See Gregory & Wallace, 1963; Gregory, 1986.
8 This description is based on a personal interview recorded with Michael John Gorman in Dublin, Ireland, in the summer of 2014.
9 See more details of the exhibition portfolio of Science Gallery Dublin (Science Gallery Dublin, n.d.).
10 The date of the remark is not certain. For details, see Jones, 1985.

Bibliography

Allora & Calzadilla & Chiang, T. (2015) The Great Silence, *e-flux Journal 56th Venice Biennale* [Online]. Available at: http://supercommunity.e-flux.com/texts/the-great-silence/ (Accessed 25 March 2018).

bobvidpix (2011) *Gyorgy Kepes "Discrepancy Between Life as It Is and Life as It Should Be"*. [Online video]. Available at: www.youtube.com/watch?v=Iie6JM2yBpA (Accessed 16 March 2018).

Cole, K.C. (2009) *Something Incredibly Wonderful Happens: Frank Oppenheimer and the World he Made Up*, Boston and New York: Houghton Mifflin Harcourt.

Daston, L. (2004) The Glass Flowers in: Daston, L. (Ed.) *Things That Talk: Object Lessons from Art and Science*, 223–254, New York: Zone Books.

Gombrich, E. (1960) *Art and Illusion: A Study in the Psychology of Pictorial Representation*, Princeton and Oxford: Princeton University Press.

Gregory, R.L. (1986) *Hands-On Science: An Introduction to the Bristol Exploratory*, London: Duckworth.

Gregory, R.L. & Gombrich, E.H. (Eds.) (1973) *Illusion in Nature and Art*, London: Duckworth.

Gregory, R.L. & Wallace, J.G. (1963) Recovery from Early Blindness: A Case Study, *Experimental Psychology Society Monograph* [Online], 2. Available at: www.richardgregory.org/papers/recovery_blind/recovery-from-early-blindness.pdf (Accessed 24 March 2018).

Jones, E.M. (1985) *'Where Is Everybody?' An Account of Fermi's Question* [Online]. Available at: www.osti.gov/accomplishments/documents/fullText/ACC0055.pdf (Accessed 25 March 2018).

Kepes, G. (1956) *The New Landscape in Art and Science*, Chicago: Paul Theobald.

Kepes, G. (1944) *Language of Vision*, Chicago: Paul Theobald.

MIT Museum. (2019a) *Past Exhibitions* [Online]. Available at: https://mitmuseum.mit.edu/explore/exhibitions/archive (Accessed 03 July 2019).

MIT Museum. (2019b) *Collections* [Online]. Available at: https://mitmuseum.mit.edu/explore/collections (Accessed 03 July 2019).

Murray, G.M. (2009) Foreword, in: Cole, K.C. (Ed.) *Something Incredibly Wonderful Happens: Frank Oppenheimer and the World he Made Up*, XIV, Boston & New York: Houghton Mifflin Harcourt.

Science Gallery Dublin. (n.d.) *Previous Exhibitions at Science Gallery* [Online]. Available at: https://dublin.sciencegallery.com/archive (Accessed 20 May 2018).

Sleigh, C. & Craske, S. (2017) Art and Science in the UK: A Brief History and Critical Reflection, *Interdisciplinary Science Reviews*, 42(4), 313–330.

Wellcome Collection. (n.d.) *Making Nature: How We See Animals* [Online]. Available at: https://wellcomecollection.org/MakingNature (Accessed 25 March 2018).

Wellcome Trust. (1996) *Press Release: Sci ~Art – Creative Partnerships Between Scientists* [Online]. Available at: https://web.archive.org/web/19970702080019/www.wellcome.ac.uk/wellcomegraphic/a4/index.html (Accessed 25 March 2018).

4 Art as a narrative tool

Seeing the unseen

When art makes its way into the curatorial discourse of science museums, it is often because of its narrative and conceptual power. Art can prove to be very effective in inspiring metaphors that can transfer a concept and stimulate reflection. Art may help us to visualise scientific objects, data or phenomena that cannot be easily displayed in an exhibition, due to their dimension, duration or nature. When we think of the aesthetic power of art, we should not think of situations in which the artist's role is necessarily and simply "to illustrate, explain or make the science accessible". Authors in this section show how applying the aesthetic dimension of art to scientific objects and concepts has profound implications connected to meaning-making and to the curators'/ researchers' understanding of the objects and concepts themselves.

Ken Arnold *et al.* recount the strategies that Medical Museion at the University of Copenhagen (Denmark) adopted to deal with the challenging topic of biomedicine. Among many other considerations, the authors rightly remind us of Dewey's understanding of art as a means that helps us not only to "notice" something, but also push our thinking beyond conventional associations and the idea of unalterable values. At the Medical Museion, the choice to investigate and exhibit an "aesthetically conceptualized presentation of scientific experimentation" ultimately supported new forms of understanding of the collections and objects. Moreover, the authors underline how a transdisciplinary attitude to curatorship has supported the inclusion of new audiences and previously ignored points of view.

In some cases, scientific objects, tools, images and products can be regarded as artworks in themselves because of their strong aesthetical or fascinating character. Moreover, the very way in which a curator or artist-curator displays an object can change its interpretation and the value assigned to it, as is effectively described by Nuno Ferrand de Almeida *et al.* in their chapter about the new museographic asset of the Hall of Biodiversity in Porto (Portugal). In this case, an artist was directly appointed with the specific task of creating a "museographic metaphor" which would "make the invisible immediately visible".

Today there are many examples of collaborations where multidisciplinary teams including artists and scientists or science communicators cooperate to generate artworks that convey a message or provide new interpretations of a concept. In his chapter, James Peto describes how artists add new points of view to collections such as the medical one at the Wellcome Collection in London (UK) and help initiate new questioning processes. When art is introduced in science communication institutions it can encourage visitors to push their reflections beyond pure facts, opening minds to new understanding. In this utilitarian frame not all artworks inspired by science works do the trick: James Peto provides practical suggestions on how to avoid the sterile inclusion or production of somewhat ineffective artworks within science exhibitions.

When introducing art in science institutions we should not forget that visitors already have personal opinions on how to approach an artwork and find a meaning in it. Yet there are other factors that can influence the interpretation and the perception of the object, first of all its context. What happens when visitors come across an artwork in a science communication setting? Hannah Redler Hawes reflects on this aspect and provides various examples of how art has been introduced in exhibitions at the Science Museum, London (UK).

Any way you look at it, in these settings art does not stand alone and it provides an additional point of view to the museographic choices of curators. This also calls for reflection on the tacit "contract" between institution and visitor. What do visitors expect? What happens when the objects they are confronted with are not what they anticipated? Does the inclusion of artistic points of view in a scientific curatorial narrative bring misunderstanding or revelation?

The case studies in this section present different ways in which institutions have attempted to describe scientific topics thorough art, combining science with digital art, installations, performative arts, interactive exhibits, new media, digital fabrication and so forth.

4.1 A house of collaboration

Investigating the intersections of art and biomedicine

Ken Arnold, Adam Bencard, Bente Vinge Pedersen, Thomas Söderqvist, Karin Tybjerg and Louise Whiteley

Introduction

Enthusiasm for the *notion* of interdisciplinarity is almost ubiquitous these days; while the *practice* of partnering across specialisms is, in and of itself, hardly noteworthy.

As art historian Martin Kemp put it back in 2011, 'art-science collaboration is becoming established as a distinct curatorial practice'. 'Ultimately', he goes on, 'as with all artworks, the artist lays down the melody while encouraging the visitors to sing their songs in their own way'. In his book on the subject, *Colliding Worlds*, Arthur Miller has gone so far as to posit the emergence of a distinct third culture in which boundaries between the two fields of creative endeavour have become thoroughly dissolved (Miller, 2014). Not denying their proliferation, Philip Ball is less convinced of the significance of these initiatives, suggesting that 'the very venture of trying to link these two spheres of human creativity is often regarded by advocates as intrinsically worthy, so that Sciart risks being received like a school play: it is the effort, not the execution, that matters'. Furthermore, he suspects that 'agonizing over the relationship has become obsolete' (Ball, 2017).[1]

Risking Ball's opprobrium, or at least his boredom, this chapter points a spotlight at one particular institution's experiments in linking the spheres of art and medical science. And drawing on our experiences, we cannot resist speculating a little about the nature of these relationships as they have manifested themselves at Medical Museion in Copenhagen from the mid-2000s up until today. We chart how these relational experiments have, cumulatively, been folded into a profound organisational shift both with regard to how the institution is perceived from the outside and to methods of investigation employed internally.

Medical Museion's art projects

Since its collections were established in 1907, Medical Museion has remained focused on medical sciences as they relate to health and wellbeing. Within those boundaries, our projects nevertheless form a fluid ecology, with roles for researchers, artists and curators being freely blurred and interconnected. The arts-oriented projects these practitioners have pursued have certainly not transformed Medical Museion into some sort of art gallery, but they have unambiguously enhanced its significance as a cultural institution. Projects have not followed a fixed formula, and have varied considerably in the kinds of practitioners involved, their working methodologies and the final outputs. We have invited artists to work with us as commentators and critics, as scholar-curators, as photographic experimenters, as visualisers, as makers and hackers, as genetic and clinical analysts, as draftsmen even. And the results of their involvement have taken the form of exhibitions, events, workshops, conferences, research papers and friendships.

The initial motivation behind Medical Museion's engagement with art/science mingling was more focused: an eagerness to move away from a classical history of medicine towards the collection and display of recent biomedicine. Artistic modes of presentation were introduced as a means to deal with the difficulties of displaying biotechnology caught between the invisibly small inner worlds of its study and the grey, indifferent looking machines employed to crack them open (e.g. Boyle & Hagmann, 2017). The museum's challenge was to communicate this science through display: to engage with these "huge invisibles" (Söderqvist, 2006; Fleming, 2007). The first step was a research seminar (*Biomedicine and Aesthetics in a Museum Context*) and public conference (*Art and Biomedicine*) both held in late summer 2007. The core proposition of the conference ran:

> The microscopic, invisible world of contemporary medicine and life sciences has enormous implications for everyone. We need to find new ways to share knowledge, understand implications and critique methods. Artists are skilled conceptualists, exploring ideas rather than simply 'explaining' them, and the Medical Museion is the crucible where biomedicine, the arts and the public meet to make this possible.
>
> (from conference advertisement)

Along with presentations from scholars and practitioners investigating the frontiers of biomedical visualisation – Steve Kurtz, Critical Arts Ensemble and James Elkins, for example – the event also showcased a world premiere performance of Jacob Kirkegaard's soundwork *Labyrinthitis*.

Artist and curator Martha Fleming brought some of the threads of those initial conversations and contemplations back together in Medical Museion's first major art and science exhibition, *Split + Splice. Fragments from the Age of Biomedicine*, which opened in 2009. Fleming led a project of collective curation drawing on the work of Museion's Biomedicine on Display research team of PhD students and postdoctoral fellows. It wove together topics in history, philosophy and sociology by applying aesthetics as a common conceptual denominator – a tool with which to reconsider themes and materials in terms of cross-cutting concepts such as 'biomedical processes and flows', 'heating and cooling', 'looking in' and 'looking out'.

After these early initiatives, the museum's research seminars, conferences and public events focused increasingly on the interrelation between biomedicine, art, aesthetics, and the senses. For example, in 2011, a *Sensuous Object Workshop* was held to discuss how we experience material objects through our senses. And *It's Not What You Think – Communicating Medical Materialities* in 2013 drew together an interdisciplinary assembly of artists, designers, museum curators, and media specialists. Temporary exhibitions were also increasingly designed with a deliberate art/aesthetics dimension, for example the *Genomic Enlightenment* installation, which displayed hundreds of fibre optic illuminated gene chips from a genomic research project to raise questions about biomedicine as an Enlightenment project. To support our efforts to bring aesthetic awareness into the museum's everyday practice, we initiated research projects on the senses, including one led by Jan Eric Olsen on touch and a PhD project by Anette Stenslund on atmospheric smell.

Medical Museion has not had artist-in-residence programmes, but in two projects artists have produced work based on 'in house' investigations. In 2010–2011 British artist Lucy Lyons was based with us as a postdoctoral fellow, spending her time using (as well as contemplating) drawing as a way of examining objects and experiences. Her findings were shared through the exhibition *Experiences of Ageing*. Another collaboration saw photo artist Nicolai Howalt work with moulages of skin diseases and lenses associated with the physician and Nobel laureate Niels Finsen, who around 1900 developed light therapy as treatment for the disfiguring effects of skin tuberculosis. Howalt literally photographed through Finsen's treatment lenses, capturing the material effects of sunlight on photographic paper in the series *Light Break* (Howalt, 2015). Both projects put the examination and active re-interpretation of our material objects at their core.

A rather different project emerged from working with *Pharmacopeia*, a UK-based collaboration between textile artist Susie Freeman and family doctor Liz Lee. In cooperation with a professor of general practice from our home Department of Public Health, Freeman and Lee produced a larger than life dress-installation titled *Femme Vitale*. Its fabric incorporates 27,774 pills and capsules of 11 different drugs woven into a flamboyantly patterned pharmaceutical frock. These represent the drugs that an actual female patient took over a decade living with metabolic syndrome. The dress provides a unique aesthetic expression for the way that pharmaceuticals are woven into patients' lives: an almost overwhelming material manifestation of a disease burden that often remains semi-hidden. The work's dramatic presence in the museum's entrance has welcomed visitors to Medical Museion since 2013. And while not trying to *explain* what metabolic syndrome is, the work does communicate some more conceptual and material aspects of metabolic research, as well as providing a point of departure for discussions about the specific condition and its drugs, and more general reflections on patient experiences.

A further departure in Medical Museion's efforts to make links between art, science and the public came in the project *Biohacking and Synthetic Biology – Do Try This at Home!* The project was part of the EU platform *Studiolab;* 'inspired by the merging of the artist's studio with the research lab to create a hybrid creative space'.[2] Medical Museion's response to this theme involved installing a biohacking lab in an exhibition room, turning it into a sort of artist's studio where radically new 'craft skills' were combined with resolutely traditional craft practices and showcased for public consumption: a space deliberately poised between active laboratory and exhibition space. Rather than drawing on 'artists', we instead hosted biohackers in order to emphasise a particular conceptual approach to synthetic biology: that is the technique of inserting genetic code into the DNA of bacteria, using them to produce proteins according to a 'programmed' code. The resulting installation (co-curated by hackers, designers, and museum curators) with home-made lab apparatus allowed visitors to participate in real experiments that further toyed with the idea of democratising biological science (Tybjerg & Whiteley, 2013; Davies *et al.*, 2015).

As with many of our curatorial experiments, we learned not only from conceptualising and setting up the biohacking studio, but also from visitor responses. The lab/installation certainly did not conform to what visitors expected from an exhibition, nor did it allow for more than sporadic material engagement with the equipment. We imagined that adopting biohacking culture would help our curatorial practices shift towards deeper engagement with the material practice of science, and it certainly expanded our appreciation of the practical implications of doing so. We also gained a deeper appreciation of the negotiations involved in collaboratively 'hacking an exhibition'; the participants had apparently similar goals, but carried with them different understandings of what communication is and is for, which had to be excavated and compared in order to resolve the differing desires surrounding the project (Tybjerg, Whiteley & Davies, in press).

The idea of making an aesthetically conceptualised presentation of scientific experimentation was also at the heart of the exhibition *Heirloom*; here the 'experiment' was driven by both scientific and artistic motives. Artist Gina Czarnecki and scientist John Hunt began with the audacious notion of growing portraits of the artist's daughters from the girls' own cells on glass casts of their faces – a three-layered representation. The production of the artworks involved exploiting novel scientific techniques in order to grow human cells in a gallery context whilst minimising cost and technical requirements, processes that might feed into John Hunt's research in regenerative medicine. The project also provided a context in which to host hands-on workshops and public debates aimed at probing the ethics of using human tissue in biomedical research, art and museum display; and the ownership and identity of such tissues. Like many bioart projects,

Heirloom experimented with the use of biological material and methods to create living artworks. At another level, it also contributed to an ongoing enquiry at Medical Museion into the history, significance and meaning of medical specimens: fragments of human bodies that through scientific innovation, craft techniques and artistic strategy have been created, collected, researched and then publicly shared (Whiteley *et al.*, 2017).

The exhibition *Mind the Gut*, which opened in October 2017, is Medical Museion's latest and most expansive experiment with co-curation and interdisciplinary entanglements, which started with *1000 Handshakes* (see case study 6.4.3 in this volume) and the accompanying exhibition *Hello Bacteria! Mind the Gut* was an 18-month long investigation of what happens when you invite artists, scientists, curators, and scholars to work together on the 'total medium' of the exhibition, and what kind of experiences and conversations about biomedical science they might create together. In order to facilitate this process, we put out an international call for 'curious, inventive collaborators' to join our team. The three artistic and two scientific co-curators began with a title (*Mind the Gut*) and a clutch of shared starting points: a concern with the interaction between gut, brain, and bacteria; a burgeoning scientific as well as popular literature on the topics; the long history of investigations into our 'gut feelings'; the common aim of creating an exhibition that meshed, science, art and medical history; and, the idea that this unusual curatorial process would itself be worthy of research. A key challenge for the project was how to retain open minds about who would do what and in what shape for long enough not to re-inscribe narrow disciplinary boundaries; balanced against the practicalities of distributing decisions and exhibition making across a motley crew with differing time commitments.

Impacts on outreach and in-reach

Undertaken as a series of experiments, these projects have affected both the external face of our institution and its inner workings; impacting on the one hand our efforts to engage the public and on the other the manner in which we frame our topics and dive into them. Most significantly, they have helped these two versions of Medical Museion's work become mutually enriching.

Externally, they have had consequences (more or less intentional) on our public presence. For one thing, they have tempted us to promote the museum's activities to a new cohort of potential visitors, the sort of people who previously routinely walked past Medical Museion on their way to the Danish Museum of Art and Design just a few doors away. We have also noticed a different type of critical attention, with notices and reviews appearing in arts and media pages and radio programmes devoted to culture, reflecting a general shift toward biomedicine becoming a legitimate topic for cultural activity. And we too began to appraise our work not just in terms of evaluation, but also as an exercise in analysis and criticism. And once word was out, artists seeking to work in partnership, or just to be in residence, started approaching us with increasing frequency. In marketing terms, this has gradually nudged our brand. So that along with our role in preserving and exploring Denmark's medical history and the medical research taking place now, Medical Museion has also gained an unmistakable place in Copenhagen's cultural landscape.

These sci-art projects have had an equally profound impact on our sense of our own ways of working, and even on our mindset. Treated sometimes as a source of material culture, sometimes as a topic of enquiry, and other times as a platform for displaying invisible aspects of bioscience, Medical Museion has throughout the last ten years also figured as a site of collaborative investigation. Along with being a public museum, we are also a research section in the University of Copenhagen's Faculty of Health and Medical Sciences, and between the two we attempt to carve out a distinctive form of publicly situated, transdisciplinary research.[3] And in this quest, our artistic co-enquirers have had a significant influence. Research has entered their projects at every stage, with the final outputs themselves becoming sites of further research, and the processes of collaboration often being a site of methodological research from the Museion perspective. Along with our steadfast goal of designing potent opportunities for the public to engage with scientific ideas and histories, we have found ourselves increasingly tempted to work beyond what is already known, foraging into ideas that have barely been thought of, and just occasionally contemplating what seems unthinkable. More tangibly, working with artists has helped us:

1 Find unexpected depths in our quest to make things visible;
2 Give voice to less-well-heard participants in the realm of health and medicine;
3 Shift focus from the outputs of medical science to the processes of arriving there.

In the first instance then, some of these projects have enabled us to expand our understanding of the potency of visualisation. Now as Philip Ball and Sian Ede point out in a recent editorial on the subject, it is important not to imagine that the artist's role is simply 'to illustrate, explain or make the science accessible. That', they insist, 'is the job of educators, explainers, journalists, illustrators' (Ball & Ede, 2017). Nonetheless, some of these projects have highlighted the philosophically nuanced power of aesthetics when applied as a mode of enquiry. This was especially evident in Martha Fleming's curatorial approach to the exhibition *Split & Splice*, in which her visual concern with our collections was manifest not so much as an exercise in disseminating a personalised taxonomy, but rather as a sophisticated form of philosophical analysis. The end goal was an exhibition that highlighted aspects of the forms and surfaces of our collections, but the journey taken to arrive at the final display unearthed new understandings of the exhibits: how they had been made, used, collected, treasured and aged.

As Svetlana Alpers observed, no matter what the subject matter and whatever their style of presentation, because museums isolate things as "object[s] of visual interest" they inevitably offer up what they show as "art". For her, this was a fundamental "museum effect" (Alpers, 1991, pp. 25–32). Alpers' insight was a specific instance of a rather broader distinction made by pragmatic philosopher John Dewey in his 1934 book *Art of Experience*, in which he distinguished between two related but different responses to visual stimuli: recognition and perception. The former, he explained 'involves no stir of the organism, no inner commotion. But an act of perception proceeds by waves that extend serially throughout the entire organism'. Alpers' museum effect, where viewers are moved by things because they are appreciated as "art", can genuinely encourage people to delve into something – contemplate what it is, means and makes us feel – rather than just briefly noticing it, and increasingly these days, snapping and sharing it, before moving on.

Dewey's problem with just *noticing* things was that it doesn't take us beyond "conventional associations": that is the 'conception that objects have fixed and unalterable values'. This, he goes on, 'is precisely the prejudice from which art emancipates us' (Dewey, 1980, p. 95; see also Arnold, 2016). For Dewey, the role of art as liberator naturally flowed from a conviction that art was inseparably linked to the rest of life. Sure, art represents a rarefied and illuminating extension of our experiences; but it is still connected to the everyday. This is how it enables 'us to forget ourselves by finding ourselves in the delight of experiencing the world about us'. The application of aesthetics in museums is therefore not just about showing things off, but rather sharing vigorous life-enhancing insights, briefly extracted from the rest of life. And, to return to our core subject, art when applied specifically to science – and particularly to its artefacts as embodiments of scientific ideas and processes – can lead to the discovery of valuable forms of strangeness that ultimately illuminate subjective insights about the rest of life.

The art of investigation

The artists we have partnered with have been highly varied. But without exception, they have approached our subjects, materials and institutional remit in the spirit of investigation. No matter what else they have done, they have all practiced an art of research. In projects like our most recent *Mind the Gut* exhibition, the blending of artistic and scientific inquisitiveness has resulted in broad yet intricate pictures of complicated subjects. And sometimes additional insight has been gained from not trying to resolve contradictory ingredients within those pictures, just as Niels Bohr (whose childhood was actually spent in our buildings) brought about profound new understandings of light by measuring and conceptualising it in terms of both particles and waves.

On some occasions, as Philip Ball points out, interactions 'between art and science are not most productively regarded as parallel efforts at comprehension that happen to use different methodologies; art has more complex roles to play, among which is a destabilization of comfortable convictions about science and technology' (Ball, 2017). Martin Kemp too is eager to warn against the notion that underneath superficial differences, 'art and science are somehow the same thing'. The association he suggests is altogether looser and subtler: 'they well up . . . from the same inner necessities to gratify our systems of perception, cognition and creation' (Kemp, 2000, p. 20). So artists may well gather and work with evidence, and undertake experiments; but likely as not they will determinably pursue "impossible experiments", fed by evidence drawn from the most unlikely, and possibly unreliable sources.

Our experience of sci-art projects has then significantly broadened our appetite for unusual approaches to enquiry. Even before embarking on a research project, an earlier, possibly more primitive stage of "searching" is necessary – subliminal moments of ruminating upon the relative fertility of different problems and hunches. One of the lessons we have learned from these initiatives is the value during this pre-research phase of, at least sometimes, avoiding topics and initiatives that are bounded by one academic tradition, favouring instead ideas that sit uncomfortably between two or more. What can then emerge are topics ripe for re-examination from fresh perspectives: ideas that move between brains and stomachs, between a common social gesture and genetic research, between the prescription of pills and living with a serious disease. However, we have also discovered that initial openness about outcomes and processes can lead to tensions and differences in expectations. In *Biohacking*, for example, both the museum and the hackers wanted to be changed, whilst having an implicit preference for the other to deliver their 'usual' expertise. While in *Mind the Gut* the ultimate decision-makers with regard to the exhibition production mostly remained the museum curators. Decisions about marketing have often proved to be a 'crunch point' for differences within a co-curatorial group to come to light; moments where it is harder to preserve a multiplicity of perspectives.

Interstitial searches for "ripe topics" have often led to projects that favour unusual types of exploration, occasionally relying on upside-down or sideways views of, for example, lenses for light treatment or the bacteria in our stomachs. This has also resulted in more space and time being devoted to listening to non-medical voices; encouraging us for example to focus on experiences of the elderly being unlocked by seeing drawings of their own hands, or extramural biological research from biohacking labs. Seen from one perspective, the resulting artworks have applied aesthetic tropes in order to establish epistemological legitimacy for non-established academic sources. This methodology echoes the emphasis placed by cognitive psychologist Howard Gardner in his landmark book *Frames of Mind* on non-traditional notions of "intelligence". For him musical, bodily, visual and other types of acumen needed to be recognised and integrated with more traditional forms of knowledge (Gardner, 1983). John Dewey's aesthetic ideas again offer philosophical insight into the specific role played by art in this context. Art, he pointed out, does not just lead people *towards* experiences in the manner of signboards, but rather genuinely *constitutes* those very experiences. (Dewey, 1980, p. 85) And indeed, we found some of our artist-led projects substantiating our nascent concerns with alternative ways of tackling medical ideas, helping us maybe move away from a tendency to over think and under feel our topics.

Working on projects that dealt with *experiences* also gave us the opportunity to dwell on processes as much downstream products – taking us back to the science, but now scrutinised as a set of assumptions and activities, a way of life even. Particularly compelling here were processes that to non-specialists might, especially when viewed through the lens of art, might appear intriguingly unusual. Thus the biohacking project rendered techniques familiar to biologists strange by virtue of being re-contextualised and shared in public. As Martin Kemp points out, there are many procedural affinities between art and science: 'observation, structured speculation, visualization, exploitation of analogy and metaphor, experimental testing, and the presentation of remade experience in particular styles' (Kemp, 2000, p. 4); so that in a project like this we were exploiting processes familiar in art to emphasise things *going on* in science rather than what *comes out* of it.

At play in *Biohacking*, and to some extent in Gina Czarnecki and John Hunt's *Heirloom* too, were the implications of artfully tinkering with biological reality through scientific tools, but for extra-scientific ends. Describing the work of SymbioticA (Australian pioneers of these sorts of experimental installations), Czech philosopher and designer Denisa Kera has highlighted the manner in which this work can hover at the edge of believably prescient fantasies: 'a way of doing science fiction in the visual arts, which coincidentally makes the fiction part closer to reality than we would expect!' (Kera, 2010, p. 221). And for practitioner and commentator Jens Hauser, the opportunities offered by this approach to bringing art and science together move the common pursuit of transdisciplinarity onto far richer territory: 'point[ing] to possible epistemological strategies where the Humanities and the Natural and Medical Sciences can merge in-depth, instead of just opportunistically looking at each other' (Hauser, 2016).

From moments of toying with unlikely scientific futures to more straightforward exercises in visualising hidden parts of medicine, our sci-art experiments have animated a whole series of pressing aesthetic, methodological, philosophical and ethical questions. But in the end, their summative effect has actually been to take us back to the fundamental concern that gave birth to museums in the first place: curiosity.

Conclusions (with a final note of caution)

Projects such as those described may help rejuvenate museums as 'houses' (Whiteley *et al.*, 2017) that promote active and collaborative enquiry: the sort of mischievous creativity, with just a hint of futility, evoked by Michel Foucault in his exalted description of a form of activity that:

> evokes the care one takes for what exists and could exist; a readiness to find strange and singular what surrounds us; a certain relentlessness to

> break up our familiarities and to regard otherwise the same things; . . . a casualness in regard to the traditional hierarchies of the important and the essential . . . I dream of a new age of curiosity.
>
> (Foucault, 1980, p. 229)

Medical Museion's foremost concern has always been with medicine and health as understood in socio-cultural contexts, and in the particular material context of the museum. And yet we have spent more than a decade repeatedly stirring the arts into our work. As the preceding pages make clear, our conclusions from surveying those partnerships are largely enthusiastic; and our intention is definitely to carry on pursuing similar future opportunities. However, we are also convinced that in order not to underplay other equally rich aspects of our subject matter and mission, a significant proportion of our future work will have completely different characters. We will not simply become a sci-art venue.

The danger of adopting an overly dominant science-meets-art modality is that other disciplines and perspectives could be side-lined, for example leaving little room for enlightenment gained from the history or sociology or ethnography of medicine. As a habitual inclination it might also lead to side-lining other more socially driven aspects of a museum's role: letting us become so obsessively concerned with the mutual illumination of art and science that we forget to explore opportunities genuinely to effect palpable differences in people's non-intellectual lives. Furthermore, an excessively sci-art-ish agenda could narrowly predetermine the nature of exhibitions and event making, tempting an institution into programming as if it were a Kunsthalle, which in terms of our public profile, might limit the accessibility of its fare. These are harsh warnings-to-self of what might result if we were tempted to become unthinkingly axiomatic about this type of work. Overall, our inclination is to proceed along this path with enthusiasm, but to temper that with gentle scepticism.

Notes

1 For a skeptical review of this idea, see Michael John Gorman's review 'The Third Culture' (Gorman, 2014). Gorman sees less a coherent movement than exciting, experimental and mutually enriching collisions. For him, the 'important question that remains is whether such art can alter the direction of scientific research, beyond provoking public discussion and debate'. Attention paid to this activity has appeared in both science journals, like the *Lancet* and *Nature* (where, for example, Kemp published a series of articles) and in the arts (e.g. Finkel, 2016; Ball, 2017).

2 See https://cordis.europa.eu/project/rcn/99496/factsheet/en (accessed 03 July 2019).

3 Concern with this type of work has brought with it a range of nomenclature: multidisciplinary, interdisciplinary, participatory and now transdisciplinary. The latter seems to be emerging as a favoured term for work that combines interdisciplinarity with participatory approaches to creating knowledge: i.e. projects that integrate academic researchers from different unrelated disciplines, non-academic participants, and the public. Examples include UCL's Population Health and Behavior

Change conference 2017, the Wellcome Centre for Cultures and Environments of Health in Exeter; Imperial College's Institute for Global Health innovation; University of Luneburg's Institute for Ethics and Transdisciplinary Sustainability Research; The New School's Transdisciplinary Design Program in New York. See Bernstein, 2015.

Project data

Art and Biomedicine (public conference) and Biomedicine and Aesthetics in a Museum Context (research workshop) (2007) Organisers: Martha Fleming; Thomas Söderqvist, Jan-Eric Olsén and Stine Hebert.

Labyrinthitis (2007). Sound art work. Artist: Jacob Kirkegaard.

Split + Splice. Fragments from the Age of Biomedicine (2009–2010). Exhibition at Medical Museion. Core Exhibition team: Martha Fleming (artist, lead curator, producer), Søren Bak-Jensen, Susanne Bauer, Hanne Jessen, Sniff Andersen Nexø, Jan-Eric Olsén (researchers and curators), Mikael Thorsted and Lars Møller Nielsen (design). The project was part of the project *Biomedicine on Display* (PI Thomas Söderqvist, funded by Novo Nordisk Foundation). Awarded with the Dibner Award for Excellence in Museum Exhibitis in 2010, a prize for exhibitions that interpret the history of science and technology.

Genomic Enlightenment (2009; revised in 2012). Installation of 650 Illumina Genotyping BeadChips. Curator: Thomas Söderqvist. Designer: Mikael Thorsted.

Experiences of Aging (2012). Exhibition. Artist: Lucy Lyons.

Biohacking: Do it Yourself! (2013–2014). Biohackers: Malthe Borch, Rüdiger Trojok and Sara Krugman. Curators: Karin Tybjerg and Louise Whiteley. Part of European Project *Studiolab*, led by Science Gallery Dublin.

Femme Vitale (2013). Installation. Art work: Pharmacopeia (Susie Freeman and Liz Lee). General practice advisor: Jakob Kragstrup. Producer: Bente Vinge Pedersen.

1000 Handshakes and *Hello Bacteria!* (2014). Performance and exhibition. Artist: Francois-Joseph Lapointe. Curators: Louise Whiteley and Adam Bencard.

Light Break (2014). Exhibition. Artist: Nicolai Howalt. Curator: Ion Meyer.

Heirloom (2016). Exhibition. Artists: Gina Czarnecki and John Hunt. Curators: Louise Whiteley, Malthe Bjerregaard, Karin Tybjerg and Bente Vinge Pedersen. Heirloom was a part of EU Creative Europe project Trust Me, I'm an Artist.

Mind the Gut (2017). Exhibition. Core exhibition group: Adam Bencard, Louise Whiteley, Nanna Gerdes, Niels Chr. Bech Vilstrup, Malthe Boye Bjerregaard, Daniel Noesgaard, Caroline Thon, Naja Ryde Ankarfeldt, Marie Balslev Backe, Christian Bache Billesbølle, Mogens Jacobsen and Guston Sondin-Kung.

Bibliography

Alpers, S. (1991) The Museum as a Way of Seeing, in: Karp, I. & Lavine, S.D. (Eds.) *Exhibiting Cultures: The Poetics and Politics of Museum Display*, London and Washington, DC: Smithsonian Institution Press.

Arnold, K. (2016) Thinking Things Through: Reviving Museum Research, *Science Museum Group Journal* [Online], Spring. Available at: http://journal.sciencemuseum.org.uk/browse/issue-05/thinking-things-through/ (Accessed 04 May 2018).

Ball, P. (2017) postARTandSCIENCE: A Symposium at the Wellcome Collection, London, *Interdisciplinary Science Reviews*, 42(4), 395–398.

Ball, P. & Ede, S. (2017) Art and Science – Work in Progress: Observations, Opportunities, Obstacles, Vol 1 (of 2), *Interdisciplinary Science Reviews* [Online], 42(4), 309–312. Available at: https://doi.org/10.1080/03080188.2017.1381221 (Accessed 04 May 2018).

Bernstein, J.H. (2015) Transdisciplinarity: A Review of Its Origins, Development, and Current Issues, *Journal of Research Practice* [Online], 11(1). Available at: http://jrp.icaap.org/index.php/jrp/article/view/510/412 (Accessed 04 May 2018).

Boyle, A. & Hagmann, J-G. (Eds.) (2017) *Challenging Collections: Approaches to the Heritage of Recent Science and Technology*, Washington, DC: Smithsonian Institution Scholarly Press.

Davies, S., Tybjerg, K., Whiteley, L. & Söderqvist, T. (2015) Co-Curation as Hacking: Biohackers in Copenhagen's Medical Museion, *Curator*, 58, 117–131.

Dewey, J. (1980) *Art as Experience*, 18th ed., New York: Perigee Books.

Finkel, J. (2016) Forget Art + French – Now It's Art + Science, *The Art Newspaper*, 275.

Fleming, M. (2007) The Huge Invisibles, *Årsskrift for Medicinsk Museion*, 4, 53–65.

Foucault, M. (1980) The Masked Philosopher, in: Rabinow, P. (Ed.) *Michel Foucault: Ethics, Subjectivity and Truth*, 1, London: Penguin.

Gardner, H. (1983) *Frames of Mind: The Theory of Multiple Intelligences*, New York: Basic Books.

Gorman, M.J. (2014) The Third Culture, *Nature*, 510(7404), 216.

Hauser, J. (2016) A New Alchemy: Biotechnology as Art, *University Post University of Copenhagen* [Online]. Available at: https://uniavisen.dk/en/a-new-alchemy-biotechnology-as-art/ (Accessed 04 May 2018).

Howalt, N. (2015) *Light Break – Photography/Light Therapy*, Copenhagen: Fabrik Books.

Kemp, M. (2011) Artists in the Lab, *Nature*, 477(7364), 278.

Kemp, M. (2000) *Visualizations: The Nature Book of Art and Science*, Berkeley: University of California Press.

Kera. (2010) The Museum as 21st Century Bestiary, in: Filippoupoliti, A. *et al.* (Eds.) *Science Exhibitions, Curation and Design*, Edinburgh: MuseumsEtc.

Miller, A.I. (2014) *Colliding Worlds: How Cutting-Edge Science Is Redefining Contemporary Art*, New York: W.W. Norton.

Söderqvist, T. (2006) Who's Afraid of the Recent Biomedical Heritage? *Opuscula Musealia*, 15, 99–105.

Tybjerg, K. & Whiteley, L. (2013) From Housewife to High Tech: Biohacking and Synthetic Biology, *Science Fiction Studies*, 40(2), 406–409.

Tybjerg, K., Whiteley, L. & Davies, S. (in press) The Life of a Hacked Gene Gun: Object Biographies in Science Communication, in: Davies, S. & Felt, U. (Eds.) *Exploring Science Communication: Using Science and Technology Studies to Research Public Communication of Science*, Sage.

Whiteley, L., Stenslund, A., Arnold, K. & Söderqvist, T. (2017) The House' as a Framing Device for Public Engagement in STEM Museums, *Museum and Society*, 15(2), 217–235.

Whiteley, L., Tybjerg, K. & Pedersen, B.V. (2017) Displaying the Researched Body: Growing Cell Portraits in a Medical Museum, *Leonardo & Leonardo Music Journal*, 50(1), 86–87.

Further readings

Arnold, K. (2017) A Very Public Affair: Art Meets Science, *Interdisciplinary Science Reviews* [Online], 42(4), 1–13. Available at: https://doi.org/10.1080/03080188.2017.1381224 (Accessed 04 May 2018).

Arnold, K. (2016) Foreword, in: Callard, F., Staines, K. & Wilkes, J. (Eds.) *The Restless Compendium: Interdisciplinary Investigations of Rest and Its Opposites*, Basingstoke: Palgrave MacMillan.

Arnold, K. (2012) A Strange Crossover, in: Ingham, K. (Ed.) *Wonder Chamber*, Cardiff: Fotogallery.

Arnold, K. (2009) Medicine and Art at Wellcome Collection, in: *Medicine and Art: Imagining a Future for Life and Love – Leonardo da Vinci, Okyo, Damien Hirst*, Tokyo: Mori Art Museum.

Bencard, A. (2019) Exhibitions as Philosophical Carpentry – on Object-Oriented Exhibition Making, in: Bjerregaard, P. (Ed.) *Exhibitions as Research: Experimental Methods in Museums*, Oxford: Berg.

Bencard, A., Whiteley, L. & Thon, C.H. (2018) Curating Experimental Entanglements, in: Hansen, M., Gregersen, A. & Folke Henningsen, A. (Eds.) *Curatorial Challenges*, London: Routledge.

Callard, F. & Fitzgerald, D. (2015) *Rethinking Interdisciplinarity Across the Social Sciences and Neurosciences*, Basingstoke: Palgrave Macmillan.

Ede, S. (2005) *Art & Science*, London: I.B. Tauris.

Ede, S. (2000) *Strange and Charmed: Science and the Contemporary Visual Arts*, Lisbon: Calouste Gulbenkian Foundation.

Mordhorst, C. (2009) The Power of Presence: The 'Cradle to Grave' Installation at the British Museum, *Museum and Society*, 7(3) 194–205.

Pedersen, B.V. & Söderqvist, T. (Eds.) (2013) Femme Vitale Catalogue, *Medical Museion* [Online]. Available at: www.museion.ku.dk/femme-vitale-webcatalogue/ (Accessed 04 May 2018).

Söderqvist, T. (2009) *Split and Splice: Fragments from the Age of Biomedicine – New Exhibition at Medical Museion* [Online]. Available at: www.museion.ku.dk/2009/06/split-and-splice-fragments-from-the-age-of-biomedicine-new-exhibition-at-medical-museion/ (Accessed 07 May 2018).

Söderqvist, T. & Pedersen, B.V. (2013) Biomedicine on Display: Copenhagen's Medical Museion, in: Alberti, S. & Hallam, E. (Eds.) *Medical Museums: Past, Present and Future*, London: Royal College of Surgeons.

Whiteley, L., Tybjerg, K., Pedersen, B.V., Bencard, A. & Arnold, K. (2017) Exhibiting Health and Medicine as Culture, *WHO Public Health Panorama*, 3(1), 59–68.

4.2 Sophia's Whale and the hypercubic showcase of sudden comprehension

Nuno Ferrand de Almeida, Nuno Valentim, Luís Mendonça, Rui Mendonça and Maria João Guimarães Fonseca

In memory of Jorge Wagensberg[1]

This contribution is dedicated to Jorge Wagensberg, the generous genius who, with unrepeatable wit and boundless wisdom, had the sole capacity of leaving us in sheer awe over the beauty of life in all of its forms. The core notions and principles that shape Jorge's vision – through which he forged the CosmoCaixa Science Museum in Barcelona as well as many successive projects – are enclosed in this chapter. The Hall of Biodiversity was in fact conceived through a deep collaboration with him, reflecting his idea of science museums as the new cathedrals, pivotal spaces of dialogue and interaction, where everyone, regardless of their past, present and future, can come together for emotional and cognitive enjoyment.

The hall of biodiversity

In 2010 the Institute of Botany "Dr. Gonçalo Sampaio" closed its doors permanently. For 60 long years it was hosted by the Andresens' House, right at the entrance of Porto's Botanical Garden, and was part of the Faculty of Sciences of the University of Porto. With brand new facilities located nearby, the old Institute finally merged with the complementary Institute of Zoology "Dr. Augusto Nobre" and gave rise to a much needed modern Department of Biology that is thriving today in the burgeoning and highly competitive fields of cancer and neurobiology sciences, marine biology, ecology and evolution, bioinformatics, amongst many others.

The Andresens' House was a magnificent 19th-century bourgeois palace that already existed in 1865, surrounded by fields sloping down to the river Douro. The views over the mouth of the river and the Atlantic were extraordinary and all these features eventually led the Andresen Family to buy the property around 1895 and live there until 1950. While the family ultimately sold the entire property to the University of Porto, at that time two of the most prominent figures of Portuguese culture – Sophia de Mello Breyner Andresen and Ruben A. – were born or raised in this magnificent building. Their novels and poems are full of references to the magical and inspiring world of the house and gardens. Today, there is not a single child in Portugal that is not familiar with *Rapaz de Bronze*, *A Floresta* or *A Menina do Mar*, all beautiful short novels that Sophia wrote for children moved by the spirit of the place.

Yet back in 2010 the Andresens' House was empty and faced an unknown destiny. In fact, it was worst: the ideas that were circulating at the University about what to do with the Andresens' House were not at all exciting. In fact, neither was the destiny of the Natural History Museum, which had been closed to visitors for more than 20 years, losing a significant part of its collections due to lack of curators, total abandonment and degraded infrastructures.

The same was happening to the Herbarium and many other scientific collections that were regarded as old fashioned and totally useless.

At that very same time, and due to an extraordinary coincidence, one of us was reading Sophia's book *Histórias da Terra e do Mar (Stories of Land and Sea)* and, in particular, one of its stories titled *A Saga (The Saga)*, which talks about a Danish sailor – Jan Heinrich Andresen (Sophia's great-grandfather) – and about how he arrived in Porto to become a successful trader and initiate a new life. It is still thrilling for us to remember that unique moment when we read Sophia (Andresen, 2015a, p. 1q15 – free translation by authors) describing the Andresens' House:

> Everything in the house was immensely large, from the bedrooms where the children rode their bicycles to the huge atrium where all the chambers opened to, and in which, as Hans said, one could assemble the skeleton of a whale that had rested for years, wrapped in numerous boxes, in the basement of the Faculty of Sciences since there was no place where it could be assembled.[2]

Was this possible, or even true? We knew that whale very well. It was the skeleton of a Blue Whale, abandoned in the then closed Natural History Museum in the Faculty of Sciences of the University of Porto. We knew what an extraordinary loss it was for the city of Porto to have been deprived of its Natural History Museum, and we knew that most of Porto citizens could no longer remember that such a magical institution had once existed in their city. We knew that children could no longer be inspired by the emotions that only a Natural History Museum can offer. Was this possible, or even true? Was Sophia offering us the solution or, at least, showing us the way? Indeed, she was.

Excited by the discovery, we quickly composed an image where the skeleton of the whale was superimposed on a beautiful photograph of the hall in the Andresens' House. The effect was indeed amazing. We called it "Sophia's Wish" and imagined that it could represent the perfect metaphor for the re(construction) of the old Natural History Museum, but one that followed a different path: one where Art could cross the path of Science, where literature and life and nature could go hand in hand like in Sophia's work, where the joy and beauty of both art and science could come together. At the end of that day, we thought we had hit the right track.

A few days later a meeting with the Dean of the University of Porto was arranged to share these thoughts and its impact was huge and above all our greatest expectations: with his full commitment, the Andresens' House had found its final destiny. We needed of course a team. First, an architect for the most important of all tasks: the rehabilitation of the old palace. Then, the artists and designers as the creative engine of the whole project. Finally, the science communicators. In a few days the whole team was in place, and we all knew we wanted to do something different, something unique, something that was at the level of Sophia's Wish. It was immediately clear for us that the Andresens' House was going to be a journey through Life, an understanding of the extraordinary and exuberant diversity of the living beings that share this planet with us. It was immediately clear for us that we were going to create The Hall of Biodiversity.

Sadly, however, we realised that a fundamental piece was missing: a museographical perspective. We were lacking the crucial expertise of someone who could understand this metaphor of Sophia, someone who could easily cross borders between science and art, someone, in fact, who could fuse both cultures and put art at the service of science. We simply did not know such a person. We thus decided to travel across Europe visiting all types of museums in Paris, London, Berlin, Brussels, Rome, and so many other places. It was not at all easy. But one day someone pointed us towards an extraordinary object: a new type of showcase used at a Science Museum in Barcelona, called the *Hypercubic Showcase of Sudden Comprehension*. We were immediately fascinated by it. But, at the same time, we also knew that we had found the missing element: we ran to Barcelona. We still have a vivid memory of the effect those *Hypercubic Showcases of Sudden Comprehension* had on all of us during our first visit to the CosmoCaixa Science Museum in Barcelona. To describe them, there are no better words than those of their inventor, Jorge Wagensberg, who was then Director of CosmoCaixa:

> If we accept that aesthetic pleasure is a delicate balance between what can be anticipated and an element of surprise, that harmony is a degree of repetition in space, and that rhythm is a degree of repetition in time, then we can say that a hypercubic showcase has the potential to create a rare beauty: that of scientific intelligibility.
>
> (Wagensberg, 2009)

That was it: scientific intelligibility emerging from the fusion of art and science. We asked for a meeting with Jorge Wagensberg at CosmoCaixa, and told him that we firmly believed he had found a new museographic language that could be understood by all sorts of publics independently of their background and that, at the same time, could provoke an unforgettable feeling of intellectual pleasure. Our conversation was long and pleasant.

Indeed, we will never forget the joy in his eyes when we slowly described the metaphor of Sophia's Whale, the Andresens' House, Sophia's poems, the Botanical Garden, the whole idea of The Hall of Biodiversity. Since that very first moment, we worked together until it was possible.

Art and science to explain life in a museum

Today we all know that life has been evolving on our planet for about four billion years through the process of "natural selection" discovered by Charles Darwin. We also know that life comprises an extraordinary collection of discrete forms that we call species, the majority of which has not yet been formally described. Furthermore, we know that we are losing species through the process of extinction at an unprecedented rate due to the ever-growing impact of human activity on the planet.

It was thus clear from the very beginning that we wanted people to enter The Hall of Biodiversity and be confronted with the urgent need to conserve biological diversity in face of the Anthropocene crisis. More specifically, we wanted visitors to immediately understand the four major arguments scientists use today to justify this need: beauty, ethics, economy and science. To do this, Sophia's poetry once again inspired a museographical solution for the architectural rehabilitation plan. In one of her most famous poems describing the house, *O Palácio (The Palace)*, she writes (Andresen, 2015b – free translation by authors):

> Its courtyard was interior was atrium
> Their balconies were inside
> Facing the centre

This poem suggested that visitors could enter Sophia's Hall at the Andresen's House, where the whale was emerging and immediately wander upstairs to the balconies where, in each corner, a *Hypercubic Showcase of Sudden Comprehension* would represent each of the four arguments mentioned above. The result was simply extraordinary: not only each *Hypercubic Showcase* has the beauty of scientific intelligibility, but the four showcases also engage in dialogue with each other along the inside balconies facing the Blue Whale imagined by Sophia, offering the intense intellectual joy of sudden comprehension.

By doing so, The Hall of Biodiversity is inviting its visitors to experience beauty, intelligence and understanding all at the same time (Grau *et al.*, 2017). More specifically, by experimenting this intense joy of sudden comprehension, visitors are ready to carry on, to absorb this new museographic language, to open up to new paths of reasoning, to finally understand the essence of life.

A simple and intelligible description of what life is can be obtained through an accumulation of diversity in size, shape and colour. In a modern museum, the harmonic display of this diversity can provoke in visitors a feeling of intellectual enlightenment. However, explaining the mechanism of natural selection is much harder because of its invisibility.

Today we know that life evolves at a very slow pace due to the accumulation of slight modifications through time in self-replicating organisms that are then exposed to natural selection. The most celebrated example of natural selection in action is the evolution of the light and dark coloured morphs of the peppered moths in the English forests before and after the Industrial Revolution, corresponding to a body of data that was collected for more than 200 years. Yes, we scientists know this very well, but how can we explain such an idea to all types of public? How can we explain an idea that is the essence of life but is, at the same time, invisible? Gravity is easy to show; natural selection is not.

If natural selection is at the heart of what life is and may be described as one of the most powerful ideas devised by the human species, the fact remains that its explanation is all but simple. At The Hall of Biodiversity, the aforementioned innovative language of scientific museography combines objects, phenomena and metaphors.

Another example is how we tell a story that is both unique to Portugal and related to the deep interest that Charles Darwin had for wild rabbits from the tiny island of Porto Santo, close to Madeira. Darwin knew that a particular weakness of his theory of evolution by the mechanism of natural selection was the lack of an experimental demonstration of how one species could be transformed into a different species. Darwin also knew that wild rabbits were introduced to the Porto Santo island by early Portuguese sailors at the beginning of the 15th century, and quickly evolved to a remarkably different form, which he described in detail in many of his writings. In a notable letter that will soon be on display at The Hall of Biodiversity, Darwin confessed to a colleague that he may have found the missing proof of his theory of evolution. Now (a hyper-realistic sculpture of) Charles Darwin sits at the Hall of Biodiversity, caressing a common European rabbit and one from the island of Porto Santo, which sit on his lap, watching five examples of phenomena that while happening too fast, too slow, too far away, on a micro-scale or behind opaque surfaces are invisible to the human eyes, reminding us of the limits of our perception. Next to him, a full-scale three-dimensional plate tells us the story of how lagomorphs evolved, depicting all the variations that prompted and were shaped along those processes. Hence, very briefly, by combining objects and museographic metaphors we are able to explain the major mechanism – even if invisible – that is at the origin of the diversity of life on our planet.

More than 10,000 years ago, humans began to slowly change their environment by inventing animal and plant domestication, eventually leading to sedentarism, the emergence of the first settlements, agriculture, cities and finally the organisation of our modern societies. The key element in this extraordinary process (that we call the Neolithic) is precisely the transformation of a wild plant into a domestic plant, of a wild animal into a domesticated animal. It relates to how humans were able to transform a wolf into a wide variety of dog breeds, to transform a wild boar into a multitude of pigs, to transform teosinte into maize. We have called this process – which is of the uttermost importance for the sustainability of our planet and the survival of our own species – "artificial selection". While this process may be more easy to understand for the general public due to its speed, which is by orders of magnitude higher than the process of natural selection, the fact is that more and more people live in cities, detached from the natural world, and understanding less and less how we have changed our environment in such a dramatic way.

Of all domestication processes that led to the daily production of those food resources that currently feed more than seven billion people on Earth, the most dramatic was surely the transformation of teosinte into maize, which originally took place on the Mexican highlands. Such an extreme phenotypic transformation was in itself a biological conundrum, which only the full genome sequencing of both wild and domesticated varieties was able to solve. At The Hall of Biodiversity we adopted this example of maize domestication for two additional reasons: first because maize is known to virtually every visitor, second because it was one of the first plants to be genetically modified, adding a third layer of complexity to our challenge in communicating this to the public through a museographic metaphor.

Again, it was an artist who provided the solution. Again, it was an artist who provided the intuition about a piece of reality that only scientists seemed to be able to understand. Again, it was an artist who was creative enough to build a rare beautiful piece of reality that made visible what was invisible. It was an artist who devised the *Corn Curtain*.

The *Corn Curtain* is an iconic exhibit: a showcase 3 metres high and 4 metres wide where four different panels are displayed. The first represents teosinte cobs, the second the extraordinary variety of maize cobs of all sizes, shapes and colours still available in Central and South American countries, the third the local varieties of Portuguese maize which were developed from specimens arrived in post Christopher Columbus times and that are already depleted of most of previous diversity created during the initial stages of domestication, and finally the fourth exhibiting strictly identical cobs that have been genetically modified.

While this unique showcase is not hypercubic, it also causes a sense of realisation and sudden comprehension. The intellectual enjoyment is there and visitors immediately understand that it was through artificial selection and agriculture that humans initiated their appropriation of the planet. They also understand that after a phase of exploration characterised by extremely high levels of diversity comes a phase of improvement, where diversity is reduced and traits are continuously ameliorated. Finally, by understanding that we live today in the century of biology, visitors also understand that highly sophisticated molecular techniques now offer unlimited possibilities to change all living beings and the biosphere as we know it.

Life is above everything the exuberance and diversity of beauty. As far as we currently know, life is exclusive of our planet and our understanding of the main mechanism leading to our current biosphere is the process of natural selection.

Life is also art in itself and since Renaissance times this has been constantly recognised. The Hall of Biodiversity is a new type of museum where art and science merge to offer visitors a permanent spectacle of intelligence and beauty, in the eternal words of Jorge Wagensberg. But The Hall of Biodiversity is also a meeting place where people from all origins and backgrounds come again and again and can think about the future, about the impact we as a species have on the planet and, most importantly, about the limits, the ethics and the relevance of finally being able to modify life and shape the tiny and fragile coating that surrounds our planet, which we call biosphere.

Final thoughts

At the Hall of Biodiversity there is an exhibit that features a life-sized artistic representation of a birch forest filled with peppered moths that automatically change their appearance, from clean to polluted, once the visitor approaches.

Museographic metaphors such as this one can make the invisible immediately visible or, in other words, make instantly understandable what scientists are still unable to explain widely today through facts and reasoning.

When thinking about the idea of bringing together in the same space art and science Jorge Wagensberg asked himself: can artists provide intuitions about reality that could otherwise only be understood by scientists? The answer is yes, as museographic metaphors can provide us with the beautiful gift of scientific intelligibility.

Notes

1 Professor Jorge Wagensberg sadly passed away on 3 March 2018, during the preparation of this book.
2 Final translation by Carla Morais Pires, a professional in this field at the Hall of Biodiversity.

Bibliography

Andresen, S.M.B. (2015a) Saga, in: *Histórias da Terra e do Mar*, Lisboa: Assírio & Alvim.

Andresen, S.M.B. (2015b) O Palácio, in: *A Obra Poética,* 662, Lisboa: Assírio & Alvim.

Grau, O., Coones, W. & Rühse, V. (Eds.) (2017) *Museum and Archive on the Move: Changing Cultural Institutions in the Digital Era*, Berlin: De Gruyter.

Wagensberg, J. (2009) The Hypercubic Showcase of Sudden Understanding (Opinion), *Museum Practice*, 47(11).

4.3 In the spirit of enquiry . . .

James Peto

As stated by William Henry Bragg, winner of the Nobel Prize for Physics, 1915: 'The important thing in science is not so much to obtain new facts as to discover new ways of thinking about them' (Koestler & Smithies, 1958).

In many science museums and science centres emphasis is placed on teaching visitors what something is and how it works – whether that 'something' is a phenomenon or an instrument. Such learning is valuable, but less so if these scientific somethings are presented purely in the context of science, simply leaving the visitor to marvel (or just nod), pocket the new knowledge and walk on to the next exhibit.

Can we make science museums feel more accommodating? I don't mean more comfortable. How can we open up opportunities for visitors to consider a variety of approaches to the subjects addressed and the ideas presented? How to create circumstances that allow visitors to connect personally to the stories being told? How to provoke a spirit of enquiry and encourage visitors to reflect on their own experiences and opinions? Should we not also be inviting speculation? Most importantly, how do we make room for – and truly value – the work of the imagination?

Faith in objects

One early morning in July 2007 a 23-year-old woman walked into the main exhibition space at the recently opened Wellcome Collection in London's Euston Rd. In a vitrine surrounded by objects and artworks from as long ago as 1100 BCE and as far away as Mexico she found her own heart, preserved in a glass jar and set into the wall of the gallery. Just 12 weeks earlier, after a long wait and with her health failing badly, Jennifer Sutton had received news from her surgeon that she could at last undergo a transplant operation to replace her failing heart whose muscles had become stiff with restrictive cardiomyopathy. A donor heart had finally been found for her.

Away from his work at Papworth Hospital her surgeon, Francis Wells, had accepted the role of curatorial advisor to the Wellcome Collection exhibition *The Heart*, an exploration of our understanding of what the heart is, what it does and what it means. Prior to surgery, Jennifer had consented to the proposal to include her explanted heart in the exhibition should a donor be found in time. Now well on the way to recovery from the operation, she described the experience of visiting *The Heart*, and seeing her own on display, as 'slightly surreal, but amazing'.

The exhibition traced the development of our anatomical knowledge of the heart, but also considered its far-reaching cultural and symbolic significance. Ancient Egyptian amulets and papyri were included, along with anatomical atlases, Aztec sacrificial tools, drawings by Leonardo da Vinci, paintings of the Sacred Heart, films by Ana Mendieta, songs by Hank Williams, folk objects, poetry, contemporary artworks by Annette Messager and Jordan Baseman, and much more.

Just across the gallery from Jennifer's heart, films, models, diagrams and news articles described the complex medical, cultural and legal obstacles faced by those who conducted the earliest experiments in heart transplantation. Nearby stood a perfusion machine, the largest object in the gallery – an ungainly agglomeration of cylinders and motors, cased in stainless steel. Should you ever have to undergo open heart surgery, a perfusion machine will take over the function of your heart and lungs while the former is disconnected (effectively 'switched off') for repair.

On its own and out of use a perfusion machine is an inelegant and inarticulate piece of scientific equipment, all tubes, wires, boxes, gauges, lights and switches. But it proved to be one of the most captivating features of the exhibition. For company it had images, objects, artworks, poems and human stories from different times and different disciplines, unified by the exhibition's exploration of why the organ that circulates our blood is so intimately bound up with the way we talk about our emotional and moral lives. Why is the heart so central to the question of the relationship between body and soul? In the context of these stories and questions the perfusion machine had a powerful impact. Not only did it provoke curiosity about its technical function, the machine also elicited an emotional response. Visitors could relate this highly technical and unprepossessing object to their own bodies and their own lives – perhaps also to their relationships with others and, indeed, with medical science.

The Heart was the opening show at Wellcome Collection. Twelve years on, its exhibition programme retains faith in the power of objects to connect with one another, to tell stories and to resonate personally with visitors, often at an emotional level. This can apply to a piece of scientific equipment or a pathological specimen as much as to a work of art, provided they are in the right company and can be understood in their broader context.

Not trying to explain

Neither strictly a science museum nor a science centre, Wellcome Collection was nonetheless conceived as a place to engage the public with science, especially medical science – past, present and future. A free museum and library, Wellcome Collection is part of Wellcome, an independent charity which funds medical research all over the world. At a time when it is so difficult for museums to survive financially we are extremely fortunate to have a single independent funder. Wellcome Collection's core subject matter, also addressed through vibrant publishing and digital programmes, is health, life and our place in this world. The Wellcome Library's rich collections offer one of the world's major resources for the study of medical history, and a programme of live events in all media tackles issues related to the social and ethical impact of developments in health and medicine.

Exhibitions typically run for four months and are often built around themes that have health and medicine at their heart, but which comfortably accommodate the interests and the work of other disciplines. By and large the subjects chosen are ones to which we hope everyone can relate: sleep, dreams, play, pain, disease, death, dirt, music, electricity, consciousness, sex, light, conflict, faith, the heart, the brain, the mind. As with all Wellcome Collection's activities, the experiences and stories of patients are featured and valued as highly as those of medical professionals. The ideas and insights of artists, writers and performers are as important as those of research scientists, philosophers and clinicians. I hope that the interdisciplinary and trans-historical approach taken helps to promote an understanding that medical science is not developed in a vacuum – that scientific knowledge changes and develops along with cultural attitudes. Cultural developments influence the path of scientific understanding, and, by the same token, new scientific knowledge often challenges our value systems and brings about widespread cultural and social change.

Though I certainly hope that our exhibitions are informative, we do not set out to explain the subjects or the science we tackle. Our aim is to provoke curiosity[1] and speculation – to engender a spirit of enquiry. Artists are important allies in this endeavour. They are integral to all areas of our programme and we work with them in a number of different ways: mounting occasional one-person shows, appointing artists to be creative consultants on temporary exhibitions, commissioning new works in response to exhibition themes, or inviting them to research and respond to aspects of our collections. The vast majority of our temporary exhibitions have included both historic and contemporary works of art.

If we argue that the scientific, the cultural, the political and the ethical are inextricably bound together, it seems only natural that artists, interested in the evolution of society, should be interested in scientific research and in how it connects to the social fabric of our world. However, when selecting existing works of art for exhibition we choose them not for their direct engagement with science, but for their individual quality and power and for their relevance to the overall subject. Engaging artworks provide alternative ways for visitors to relate to the content of the show and greatly enrich the conversations between exhibits that will bring an exhibition to life.

Artists are sometimes understandably wary of showing work, originally conceived for a neutral space, in the busy environment of a multidisciplinary exhibition. They may legitimately feel that the interpretation of their work risks being skewed by the particular context of a thematic exhibition. More often, however, artists have found it rewarding to see people find unexpected ways of looking at their work, and enjoyed seeing new significances emerge.

Sifting through the grey

Rather than feeling the need for exhibitions to 'teach' or impart expertise, it seems much more profitable to awaken and to embrace the spirit of investigation. If you want to intrigue your visitors, to encourage a questioning spirit, artists can be valuable partners. Science is still often misconstrued as a process of determining black from white, when so much of it actually involves sifting through the grey. In whatever field of knowledge they are exploring, artists tend to be drawn to the grey, to those areas that are open to interpretation, where prevailing views can be challenged, where there are mysteries or uncertainties, where the truth is disputable and the ethics difficult or unclear.

In these areas there is room for speculation, for individual interpretation, and above all for the imagination. Science – perhaps medical science in particular – is full of them, and they provide enormously fertile ground for the curious. What happens in our brains and bodies when we are asleep; the complexities of relationships between humans and animals; the edges of consciousness; enhancement of the human body and mind – these are examples of subject areas that have provided rich seams for our programmes to mine and for the artists with whom we work to wrestle with. They are subjects of huge interest and importance to science but where so much remains unknown, and where ethical difficulties wait around every corner.

Artists can play a vital role in helping us to feel our way through such complexities, giving glimpses into ways of thinking that might be different from our own, or triggering the kind of uniquely personal association that can spark engagement with an unfamiliar subject.

But let's not simply think of artists as being there to make the emotional connections – to provide feeling. If the artist is simply seen as a humanising force, bringing warmth and spirit, what does that say about the science? Science itself is nothing if not a human endeavour.

In 2008 Wellcome Collection worked with the osteoarcheologists at the Museum of London to devise the exhibition *Skeletons: London's buried bones*. The Museum of London houses an astonishing collection of over 20,000 human skeletons, most uncovered by the digging of the foundations of 20th- and 21st-century London.

The exhibition showed 26 complete skeletons spanning 16 centuries, each respectfully laid out in its own case. Around the walls were ten light-box-mounted photographs, commissioned from the artist Thomas Adank. They showed each of the London locations in which the skeletons were discovered as they are today – now occupied by an anonymous housing project, a prestigious office block or a car park. Each skeleton was accompanied by a label and diagram noting where it had been found and outlining any pathologies, injuries or diseases visible on the bones. Through painstaking forensic analysis by the team of osteoarcheologists, each skeleton reveals its own story, allowing fascinating insights into the life and times of its owner and the health hazards of the day – from syphilis to smallpox to rickets caused by malnutrition.

Resisting the temptation to flesh out the exhibition by adding didactic texts, objects or images from the places and times that the skeletons would have inhabited, the presentation relied instead on its two most basic materials: the bones of the individuals and the scientific and archaeological knowledge applied to them. To ensure that the exhibition was respectful and engaging, careful choices were made collaboratively between the curator, Emily Sargent, the osteoarcheologists and the exhibition designers about which skeletons to display and how. Reminding us that in London the dead are all around, Thomas Adank's lightboxes invited present-day Londoners to connect their city's present with its past, and to connect their own lives with the lives of those who have preceded them. But one of the great strengths of this project was that it was the *scientific* research that brought out the humanity from the skeletons of the individuals included, revealing information about their lives and health that has, until now, remained hidden, sometimes even from those who once called the bones their own.

Though the exhibition was aimed primarily at adults, it proved hugely popular with family groups. The gallery was constantly alive with conversations speculating about the experiences of the people featured. Debates about the value of human remains to scientific research, and arguments over the ethics of displaying human remains in museums are sometimes portrayed as setting scientists and humanists at odds. One of the strengths of this project is that it cut through such stereotypes. A smaller version of the exhibition continues to tour, with skeletons local to the places to which it travels adding their particular individual stories.

Finding fertile ground

As well as avoiding the risk of simply looking to artists to provide the human connection, it is also, I think, limiting to invite the involvement of an artist in a project in the hope that they will contribute something 'about' the science. Of course you want them to engage with it, but work 'about' it is less likely to be interesting. Furthermore there is a danger of 'Science and Art', or 'Sci-Art', being perceived as a discipline unto itself, which risks closing down the possibilities rather than allowing for or creating a wider range of ways of working. More important than Sci-Art is good art, and as Sian

Ede (herself a driving force behind the bold and important initiatives originally set up under the Sci-Art banner) wrote at the outset, 'Good artwork is always more than the sum of its parts and operates like poetry, making suggestions, hinting at associations, teasing emotionally, challenging intellectually and expecting the viewer to play a part in making new meanings from it' (Ede, 2005). Seeing Science and Art as some kind of specialism risks leading it down a narrow alley.

Commissioning institutions can help minimise the above risks by seeking out new partners on the basis of their approach to making art, rather than for their established engagement with science. Ideally the commission brief will allow time and space for finding the most fertile common ground between the artist's interests and the work of the institution. Time may sound like a luxury, and a potentially expensive one at that, but it doesn't have to be. Certainly the artist needs to be paid properly for their work, but models such as research fellowships, whether formal or informal, might allow for a more extended engagement, with more time for thinking, adjusting and planning on both sides.

A longer timespan and lack of pressure for an immediate outcome can work particularly well for those institutions which have collections and which enjoy encouraging artists to work with them. Collections – like the people who put them together, those who look after them, and those for whom they are being looked after – are multifaceted. They can take a while to get to know and to understand. Deeper relationships need time to develop. A longer-term commitment to an artist does not have to mean finding and funding a solid block of time. The artist-in-residence model is not the only one. It could involve a more intermittent relationship over an extended period, allowing time for discovery, experimentation and further research – time for ideas to evolve alongside other projects.

The work of the British artist Shona Illingworth has benefitted from a deep engagement with scientists over an extended period, without the science ever becoming the subject of the work. Her relationship with Wellcome Collection, too, goes back several years in which she has made significant contributions to two of our major exhibitions War and Medicine and States of Mind: tracing the edges of consciousness. It is over 14 years since she first began collaborating with the neuropsychologist Martin A. Conway. Their mutual interests in the neural complexities of memory have informed a number of collaboratively made films and installations that have looked at the intersections between individual and collective (or 'cultural') memory.

In describing the benefits of Illingworth's drive and curiosity, Conway says: 'She asked questions about memory that perhaps scientists don't normally ask, and the results of that can be seen in our work' (Wigmore, 2017) The continuing collaboration, sometimes informal, sometimes supported by research and development funding, has also introduced the artist to colleagues and patients of Conway's. Their partnerships inform the major three-screen work *Lesions in the Landscape: Claire and the Island of Hirte*

(2015). As a result of a brain trauma, Claire, a patient of Conway and his colleague Dr Catherine Loveday, can no longer remember most of her past. Claire's experience of the impact of her amnesia is central to the work. Its principal visual focus is on the landscape of the remote Scottish island of St Kilda, whose declining population was evacuated on a single day in 1930. The multi-layered soundtrack interweaves the voice of Claire with memories, sounds, imaginings and present-day images from St Kilda. Her descriptions of the experience of losing her memory are interwoven with both lyrical and jarring glimpses into the myths and realities of the remote island, suggesting a kind of cultural amnesia running in parallel with the loss of Claire's own past.

Of Claire, Conway, Illingworth and Loveday's joint engagement with the dynamic complexities of human memory, Loveday writes: 'We all have some expertise, but only Claire really knows what having amnesia feels like. I really hope Shona's work will help people to connect with the experience of amnesia in a completely different way' (Rhodes, 2015). Describing how the experience of the collaboration has changed the way in which she works with Claire clinically, Loveday adds:

> I've always seen myself as a quantitative scientist. I want numbers and accuracy. Art is as qualitative as it gets. But drawing away from my way of seeing has been a gift. With this new perspective I can go back in and ask my qualitative questions. I understand more about amnesia now, not only from an academic, clinical, or patient level, but more about what it is like to be that person.

When looking for artists with whom to work, there is a temptation for museums to turn to those who already have a reassuring history of museum collaboration or engagement with science. At Wellcome Collection we have often shown the work of artists already engaged with the subject matter of the exhibitions we are putting together. However, when commissioning new work we try to look in different places – to establish collaborations which will bring something fresh and surprising for the artist, the exhibition and our audience. This has led us to commission writers, filmmakers and performers as well as artists. The most successful partnerships began with conversations about the subject matter of a future exhibition, long before the shape of the exhibition was determined.

Would you mind?

Curated by Kate Forde and Honor Beddard, The Institute of Sexology was a 2014 Wellcome Collection exhibition about the history of the study of Sex. It looked at the work of pioneering Sexologists like Alfred Kinsey, Marie Stopes and Masters and Johnson and at how important it has been for our health and wellbeing to understand attitudes towards sex and people's sexual behaviours. The NATSAL survey (National Survey of Sexual Attitudes and Lifestyles) of the early 1980s, for example, which Margaret Thatcher refused to fund, proved enormously helpful in understanding and trying to prevent the spread of AIDS/HIV.

The work of the sexologists was frequently underpinned by surveys and questionnaires – encouraging and allowing people to speak openly and honestly. When writer, theatre director and performer Neil Bartlett was commissioned to make a work for the exhibition, he devised his own sex survey, graphically designed to look the part, with its own envelope printed with the request: *Would you Mind?* – the title of the project. A large table with seating was built into the exhibition's final section where visitors could fill out the questionnaire anonymously. A post box was incorporated, as well as a regularly updated LED screen for playing back some of the responses and the statistics gleaned.

The survey was both serious and humorous. Visitors engaged with it enthusiastically. It took an average of 20 minutes to fill in. The exhibition ran for ten months and had over 200,000 visitors – 23,000 of whom took the time and trouble to complete the 25 questions, examples of which included: 'Have your desires changed as you have got older?' 'Do you think having better sex will make you a better person?'

The questionnaire had to be reprinted regularly, but that was part of Bartlett's conceit: the last question in the survey read: 'If you could ask the other people coming to this exhibition one question of your own about sex, what would that question be?' With each reprint, one of the visitors' questions was added. By the end of the show most of the questions had been suggested by visitors. The 23,000 completed questionnaires have now been accessioned into the collection of the Wellcome Library.

The extent of the scientific value of the 'survey' is open to challenge. It was not intended to be rigorous in its methodology. There will also be those who question whether it is art. But did it add value to an exhibition about the history of the social and scientific study of sex? In my view, hugely: 23,000 people spent a considerable time reflecting on it – pondering, discussing and writing about things that are fundamental to their sense of self and how they fit into the wider world. And their anonymous thoughts will be available for future generations to look back on.

Uncertainty

Finally I would like to make a gentle plea for the value of uncertainty. This may be partly a reflection of my own position. My background is not in science or medicine, but I find myself working in a science-focused institution. That does bring a level of discomfort. However, some of the most absorbing and exciting projects I've been fortunate enough to be involved with have started from a position of uncertainty or 'not knowing': uncertainty perhaps, when setting out to curate an exhibition as to what it should be aiming to communicate, or uncertainty on the part of an artist embarking on a new commission about what he or she is going to make. Wellcome Collection exhibitions, events and commissions often feel like investigations from the outset. Those making them are rarely experts in the subject when they start out, but they talk to many experts along the way. In talking to people who approach the topic from very different perspectives, curators or artists are piecing things together as they go. The spirit of enquiry is alive through the whole process, with the result that the excitement of researching, learning and discovering unexpected connections translates into the final outcome.

I am often greatly impressed by the openness of our visitors and the extent to which they are game to join in. Neil Bartlett's project was testament to this enthusiasm, but it shows itself particularly clearly at live events in which the audience is invited to contribute or participate. Their willingness is remarkable. This may in part be a reflection of our subject area (health and the 'Human Condition' provide a great deal of common ground for sharing), but I hope their intrepidity also reflects an empowering sense on their part that their curiosity and openness is shared by the museum itself. I would also like to think that they are comfortable engaging with both science and art without worrying or even thinking about which is which.

Note

1 It is important to distinguish the use of the word 'curiosity' as applied to a state of mind (wanting to know more) from its use as applied to an object: a curiosity, or a curio. The latter are not helpful terms when describing objects in museum collections.

Bibliography

Ede, S. (2005) *Art and Science*, London: I.B Tauris.

Koestler, A. & Smithies, J.R. (1958) Beyond Reductionism, in: Daintith, J. (2009) (Ed.) *Biographical Encyclopedia of Scientists*, 3rd ed., Boca Raton: CRC Press.

Rhodes, E. (2015) Lesions in the Landscape, *The Psychologist* [Online] July, 28, 522–527. Available at: https://thepsychologist.bps.org.uk/volume-28/july-2015/lesions-landscape (Accessed 07 June 2018).

Wigmore, G. (2017) *The Art and Science of Memory, News from City University of London* [Online] June. Available at: www.city.ac.uk/news/2017/june/the-art-and-science-of-memory (Accessed 07 June 2018).

4.4 Context, collaboration and contemporary culture

Hannah Redler Hawes

Museums have always been places which, in different configurations, bring together people, ideas, actions and things. In Western culture the dominant museums of the 19th and 20th centuries were predominantly structured around subject-matter specialisms. But curious minds, broader culture and equally objects in collections are not so obedient. My curatorial interest is in constant slippage between disciplines and in involving artists, other researchers and audiences in joint journeys of discovery. My work involves constant questions around what happens when you take art outside of the art gallery, mix science and art histories and related discourses, and consequently disrupt the rules of engagement with museum visitors through interdisciplinary offers.

In this chapter I share some of the projects we programmed as part of Science Museum Arts Programme (SMAP), a contemporary art programme I ran between 1999 and 2014.[1] My observations also reflect my lessons learnt with other science, media and art museums as well as science and technology institutes. I will consider some of the opportunities, constraints and impacts on visitors and musicological practice of various approaches including commissioned and purchased interventions for science-themed galleries, self-contained art exhibitions in science settings, participatory projects, artists' research and interdisciplinary events. I look at issues around selection, presentation, interpretation, and visitor responses.

Why bring art into science settings?

The most common reason non-art organisations like to bring in artists is to offer a new and different way of thinking. Artists are continually commenting on, reflecting and making sense of changes in culture and society. Working with living artists brings in unique contemporary angles, particularly for museums whose collections are predominantly historical but who want to demonstrate their relevance to the present day. We are also in a world where not all major developments in science can be represented by iconic objects, where advances happen at the microscale of bits and bytes or the industrial scale of industry, both of which are difficult to display. Artists bring the multisensory qualities necessary to enliven exhibitions. Their contributions can also result in new ways of engaging with audiences.

Artists enjoy the opportunities to work within new contexts which locate them within unusual fields of reference, test their ideas beyond the art world, put different demands on their work and expose it to new audiences. They are also often subject matter experts, particularly in the fields of media art. For artists who have been researching a specific scientific topic, it is rewarding to be able to show the work within a context where the topic is more widely discussed and understood. Artists do not like seeing their work instrumentalised, which can occur when it is 'applied' in the service of the science messaging to the detriment of its own conceptual basis. This approach is also pretty poor art curation.

Art at the Science Museum

The Science Museum's own collections have always included a depth of visual culture (initially held across a range of collections), including photography, painting, sculpture, printmaking, drawing and craft objects. These were collected for many reasons: as examples of what the 'apparatus' could do, to portray senior figures or developments in science, to represent different ethnographic or folk practices, or as visual works of science in their own right, to list a few examples (Jardin, 2014). Although the way the museum works with art is constantly evolving, it has not always discussed its art collections in the context of wider arts-specific histories, which was perhaps a unique ambition of SMAP. The origins of SMAP go back to 1995, when the senior management team,[2] commissioned an Arts Policy (unpublished) which commits the museum to engaging contemporary artists in every major capital project (Cumberlidge, 1996). Key Policy stated aims, which in my experience are shared to a greater or lesser extent by every non-arts organisation I've worked with, are to:

- Enable the public to explore science and technology through art;
- Offer opportunities for artists to respond to the collections and to have access to curatorial and scientific expertise;
- Create innovative environments which will heighten the public's perception of both art and science and their interaction;
- Create a richer visual environment and emotional connections to stories.

The Wellcome Wing: case studies of art interventions

Conceived as a "theatre of contemporary science", with an architecture aimed to amaze and engender a mood of heightened expectation, the Wellcome Wing opened in 2000. It supported a world view informed by multiple voices and subjective opinion. For the museum, which alongside many other international museums, was exploring how to redraw more instructional modes of communicating with visitors to create more reciprocal relationships, it provided the opportunity to deploy new strategies for more discursive two-way interpretation, through four floors of landmark galleries. Art was one part of the broad interdisciplinary palette, where highly crafted content-rich messaging sat alongside the objects as well as outstanding quite experimental design and innovative interactives.

Despite this internal vision, and the fact that by 2014 they welcomed new approaches,[3] visitors did not and still do not *expect* to encounter art in the Science Museum.[4] They like the idea of artworks as "a wild card, outside the control of science" but do not always *get* or *like* it. This affects how they respond to it, ranging from what they recognise as art, to the mindset they bring to the works and the labels. Even art lovers are often unprepared to have the kind of interaction they normally associate with art galleries, such as being confounded, in a science context.

It was hoped that contemporary art in the Wellcome Wing might offer ways of prising open 'gaps' in its narratives, encouraging visitors to challenge the assumptions of science and ask their own questions.

On the ground floor four artworks joined a series of objects named *Talking Points* and were selected against a unifying criterion, that they would be "quick to get, hard to forget". Three involved viewers in considering the significant moral quandaries new developments in science raise. Yinka Shonibare's *effective, defective, creative* 2000 raises questions around on what basis people might choose to terminate a human life. It remains one of the most emotive artworks on display, not least because visitors frequently read it as a literal assertion rather than a speculative question.[5]

Digitopolis on the second floor included a major series of software driven interactive artworks commissioned against the criteria they should be outstanding examples of digital technology, immediately engaging, visually stunning and conceptually robust enough to sustain in-depth scholarly research. They were commissioned and selected as artworks in their own right, but as is often the case with media art, clearly also showcased unusual creative applications, albeit not at the expense of the artist's conceptual agendas.

Who Am I? (2000, updated 2010) on the first floor explores human identity through the biomedical themes of ageing, communication, consciousness, genetics, gender and brain science. In 2000 the selection criteria we built looked for artists whose work would critically connect to these themes whilst showing traces of the mark of the human hand to reinforce human uniqueness. We predominantly looked for works made with traditional materials; paint, pen and ink, cast iron, the human voice or pieces that suggested the unquantifiable inner mysteries of the human animal – a direct foil to the search for reliable evidence in science. Our selections were based on works we felt would individually either disrupt, reinforce or bring new questions to each case narrative, or the overall theme.

The *What's so special about human communication?* case was about how sophisticated language was deemed to be a particular human trait. The objects were unremarkable. We commissioned David Shrigley in the hope his maverick humour might 'save' the case. He created a large trophy cup engraved with its title: *The British Swearing Association Challenge Cup*. It surprised people by seemingly rewarding bad language. It inspired the designers, who were struggling with a good design metaphor, to come up with a series of attractive slender plinths of different heights to support both it and the other objects. But we still felt the case needed more "oomph". Following conversations Shrigley also created a scattered installation of objects, such as a vinyl record labelled *A terribly long and boring conversation* and texts on scraps of paper which included succinct messages such as 'Your mother called. She sounded angry' and brought the content messages to life through his laugh-out-loud funny yet sharply insightful play with the elasticity of language, intention and meaning.

Marlene Dumas's work took on a more quietly subversive role. She submitted two portraits from her ongoing *Rejects* series for a case about cloning, at the time a subject of much public fear and distrust. She wrote in pencil on a portrait of an individual we can't directly tell the gender or ethnicity of, *The Experiment*, and on a portrait, which looks like a stereotypical white male boffin, *The Expert*. She also wrote her own labels for her works, which undermined our assumed certainties of science even further, particularly by posing specific questions which she answered herself in the label text with the cryptic response 'we do not know'.

For the 2010 *Who Am I?* upgrade Dryden Goodwin created a series of works in which he attempted to discover insights into 'unknowable' strangers using portraiture in a case about brain science, which displays apparatus invented by science to try to unravel the mysteries of consciousness. Revital Cohen made *Disclosure Case* 2010, an imaginary object which speculates what we might do with the knowledge genetic information provides us with. The work referred to what was then very esoteric research looking into the potential for gold to cure cancers. Staff were nervous about including it in case visitors thought it was 'true'. It was placed in a case alongside a DNA sequencing machine which, though capable of performing amazing data mapping tasks, actually looked like a photocopier.

Some of the most meaningful responses to gallery interventions came from visitors who had been given personal tours. Reactions to art identified via labels alone were mixed (Fisher & Paynter, 2011). In the next section I look at how visitor perceptions and experiences of the art are managed.

Curating space for art

Commissions need to satisfy competing constituencies of interest: the Museum, existing audiences, funders, wider arts worlds, new audiences and exhibiting artists and museum teams. Our intentions for how the art would function were multifold: we wanted to select powerful, original works of contemporary art that resonated with but were different to the content while simultaneously creating coherent 'art trails', that would bring new thinking through their own conceptually rigorous sub-narratives.

A significant difference of bringing art into science vs art contexts is the physical and intellectual 'real estate' afforded to the art. It is unusual for the space required by the art to lead the narrative or exhibition design. In the case of gallery interventions it is quite frequent, sometimes for funding reasons and sometimes due to the development arc of the narrative, for art to be considered at a later stage than lead content objects. Elaborate exhibition design whilst creating exciting otherworldly atmospheres also exerts a significant degree of control on the material, physical and conceptual directions the artworks might take. Content teams will often be concerned about what impact 'non-messaging' art works will have on their carefully crafted messages. The location and availability of interpretation materials is essential. People encountering art interventions want to understand why it is there (Fisher & Paynter, 2011). If they can't find information they get frustrated and demotivated. The ricochet between desire for more information and resistance to reading it challenges curators to imagine new means of opening up conversations – which is an ongoing battle – while creating the perfect 'art in a non-art context' interpretation strategy is an ongoing mission for me and one I have not yet cracked.

Location matters

The artworks for *Digitopolis* had to be realised within fixed steel and glass structures which traversed the width of the gallery in five strips, their undulations creating covered and exposed areas for exhibits and art. Their form restricted the art but created a parity between it and other objects and interactives that was interesting. One of the artists mused that once the project was completed and installed he would stop seeing it as only 'his' work. He knew that it would be experienced within the exhibition design, itself within the context of the uber-brand of the museum. His observation highlighted the fact that the disruption of art interventions works both ways. Labels were limited in word count and predominantly conveyed how to get involved. There was little space to discuss the artists' broader concerns.

In *Who am I?* 2000, the budget for art came through late, when cases and content were fixed. Antony Gormley was unconvinced about placing *Iron Baby* 1999, his tiny life-sized sculpture of a new-born, into a case amongst objects. It didn't allow for enough negative space around the work and thus diminished it. At his insistence we fixed it to the open floor creating a startling prospect for visitors whose responses were equally surprising.

Toddlers lay with it, stroking and kissing it. Adults, shocked to find it on the floor, cautiously kneeled to stroke it. Teenagers in groups took turns running up and kicking it. As mobile phones became more prevalent this action 'evolved' into happy slapping – the practice of filming acts of violence on a mobile phone. The artist arrested an initial desire to remove it, concluding that he was happy it provoked strong reactions.

For *Energy* (2005–2018), we knew the target audience found the subject boring so we needed to create something sophisticated and fun to attract them. We developed content, design and artists' briefs simultaneously, creating a feedback loop between all three which resulted in a solution which took the locations and physical space required by the artists' concepts into consideration. Designers Casson Mann postioned all exhibits along tangents around a central startling artwork which delivered a (safe) electric shock as part of its experience.

With the *atmosphere* gallery (2010), contentious content at a politically sensitive time made it difficult to negotiate physical space for the art, predominantly due to concerns about the level of disruption to nuanced messaging. Superflux, the selected artists' good-humoured response was a participatory performance *A Cockroach Tour of the Science Museum*. Visitors joined lead 'roach' actors in insect costume 'shells' to explore the history of our changing climate through iconic objects, from the perspective of roaches apparently baffled at human's need for enhanced speed and power. Assistant SMAP curator Ruth Fenton proposed replacing ambitions for a permanent work for the gallery with a series of temporary projects. Outcomes included the first novel commissioned by the Science Museum from former writer-in-residence Tony White (White, 2013).

Many galleries position artworks at the beginning or end of the visitor experience. This can act as a beacon or scene-setter for the gallery, its subject or its underlying ethos, and can avoid the clashing of contexts. But it can be ignored by visitors anxious to get started or simply get out! Where the artwork sits in the visitor journey, what the likely flow and dwell around it might be in relation to the rest of the gallery should be considered as of equal importance to any other object.

The way you tell 'em

When delivering art to unexpecting audiences who have come to experience science we have broken the rules of the game and we need to provide clues as to how to now engage. The most direct means are through label texts, tours or workshops and the external communications strategy, although judging what to explain can be problematic.

In *Who Am I?* 2000 object labels coloured yellow and tagged "contemporary art", to distinguish them were allocated the same 40-words limit as the main narrative objects. This was initially the only place the art was flagged.

It allowed for a one-liner about the content or concept of each piece. Trying to offer further direction, we added the strapline "*The art in Who Am I? offers new perspectives and invites questions*". We fatally ignored that to ask questions visitors need to feel more informed. Post-opening, we considered which questions we should aim to ask *for* visitors and came up with: What am I looking at? Who made this? Why did they make it (contextualising their practice)? What was the artist thinking? What questions does it raise? The ways and layers through which we answered these questions we agreed should be tailored to the project, its target audience, budget, physical space on-gallery and scale of artworks. Digital art labels can link question hotspots to their own chunks of text.[6] On-gallery digital labels lend themselves to sustained group browsing. They also have scope for representing what other visitors think as prompts and could even include signposting to other references.

Labels can be sites of potential linguistic, attitudinal, perceptual or cognitive conflict. Recommendations include writing in more accessible language, mentioning words like "artist" or listing "art" materials, reproducing an image of the artwork on the label, positioning the label as close to the artwork as possible, and responding to the "literal" mindset visitors commonly bring to science museums by contextualising the artwork better in relation to the gallery topic (Davis, 2009).

Even when all these recommendations are met, visitor responses sometimes remain lukewarm, leading to the question: what would make an art label universally accessible? I think the problem in answering this is rooted in the distinct languages and discourses of art and science. Visitors' attempts to create meaning and make sense of the unexpected, be these aided, abetted or even hindered by our interpretation devices are inevitably open to misinterpretation. Combining the languages of art and science, be they visual, written or otherwise continues to feel 'unchartered'. Readings and references frequently become scrambled. In an ideal situation this is the grit that creates the pearl: new interpretations will be formed, new ideas catalysed and new knowledge gained. But it takes a lot of work and generosity from all parties, the artists, the curators and organisations, and the audiences.

I wonder how big an ask it is to invite visitors to think differently through art when they are already being called upon to think differently about science? What curators think of as a change to our museum environment is not a change for the visitors who have never been there. Is finding the right 'mindset' even more challenging because unless visitors already recognise the art, the otherworldly nature of high-concept narrative design environments (which the art helps create) makes it difficult to distinguish it? I do not know and neither do I think these challenges suggest refraining from interventions. We don't want to lose the element of surprise we have programmed the art to create. I remain in two minds over how much it matters that everyone recognises the art as art vs something interesting, thought provoking and appealing. But we do need to work harder. Marketing and communications strategies can have a major role to play, particularly cross-media.[7] By the time people encounter labels they are already in the gallery so being briefed in advance via pre-visit campaigns can help prepare expectations and build knowledge.

Framed isolation: a specific site for art

How is art experienced in a dedicated arts space within a science setting? The *Who Am I?* evaluation pondered whether art interventions would be more deeply engaged with as art if they were more saliently exhibited. Our temporary solo art exhibitions provided an opportunity for visitors to consider science within artistic contexts.

HEXEN 2.0 by Suzanne Treister in 2012 presented her unique critical overview of modern intellectual and scientific history through "interpretations of alternative knowledge creation systems". Works included drawings, photo-text works, a set of re-imagined Tarot cards, and a film.

Exit surveys suggested that most visitors understood it as an art exhibition because it looked like one. They responded positively because it was visually attractive, made interesting links to history, science and art and was different to what they would expect to find at the museum. Some visitors and staff, invested in the Science Museum's function to portray science accurately, found the show difficult. They disliked the artists' methodologies and her presentation of "information" through "occult" forms. Most liked the depth of Treister's research and the ways she made different connections between people, places and technological events. Comments (Harvey, 2012) included:

- 'Unbelievably well researched and terrifying . . . . It had quite an emotional effect. I liked the fact that it was genuine'.
- 'The film is inaccessible'.
- 'Forward thinking. Requires further Googling'.
- 'I don't know what 'art' means but there are lots of ideas to engage with'.

Redrawing relationships with audiences through art

From my perspective, which is one of a professional observer rather than formal researcher, the exciting shifts and 'new' museological approaches seen in expanded learning programmes, and new forms of exhibition-making through consultation and co-curation, owe much and have much to gain from artists' strategies for engagement and empowered co-production, as well as owing a debt to the pioneers of community art and socially engaged practice. In 2007 the SMAP team had the opportunity to combine the ideas of artists working in these traditions with museum learning and engagement teams, who were working on a new vision for expanded relationships with visitors.

We were briefed with a particular wider-museum curatorial challenge: the quandary of not being able to interpret developments in fast-changing digital technologies fast enough with visitors. We were struck by the way in which the art commissions for *Digitopolis* were the only exhibits which had not dated in the exhibition's five-year life-span. This reinforced the fundamental role of artists' critical voices in the field. We developed a proposal for working with community and media artists, with specialism in new democratic digital tools and collaborative ways of working to create an environment that would combine touchstone influential art works with access to hands-on, process-based ways of working. Visitors would be able to witness the probing questions artists were asking about new technologies and their impacts on our lives through the installation works, and to bring their own questions to the fore through direct interaction with artists, engineers and other participants.

This was an unusual programme for a national museum to undertake at the time. A large-scale version of the project proved difficult to fund, but with support from the Art Fund we acquired Mark Hansen and Ben Rubin's landmark media artwork, *Listening Post* 2002, created out of thousands of real-time, uncensored, text fragments culled from live online public forums. With funding from the Arts Council of England we commissioned a participatory minifestival, *Being Connected*, that looked at the concept of networking as social activity. It promoted the view of networked systems as being inclusive of people, things, ideas, hardware, software and virtual and real environments. We wanted to offer participants opportunities to develop new skills with new tools.

Working through Vital Regeneration the *London Tigers*, an all-female community group aged between 16–19 years, was recruited from the Marylebone Bangladeshi Society. They were an under-represented group within the museum's audiences whom we wanted to better connect with.

A lead artist (Graham Peet) was appointed to be a central interface between the museum, the participants and an additional group of project artists. His long-term experience of delivering meaningful workshop experiences with exhibitable outcomes was crucial. His purpose was to help bring together artistic processes and informal museum learning objectives, and to help identify the new methods we anticipated would be developed through this exploratory programme.

Four artist groups and Graham contributed a one-day workshop each within a half-term period. The artists were briefed to share their creative and technical skills, empower the participants and enable production of a high-quality collaborative work for public display. We hoped the participants would discover that, as users or future professionals, they have creative opportunities, an active role to play in the development of software and hardware, and in the social and political ramifications these embody.

MediaShed worked with the Tigers using devices to identify hidden wireless CCTV cameras to perform in front of, later accessing the captured images to create their own film; Furtherfield and the group mixed audiovisual material to create short film collages that reflected on life in a hyperconnected world; Jonah-Brucker Cohen and Katherine Moriwaki provided discarded junk and equipment to make robots and electronic musical instruments, while Graham worked with the group to assemble photographic portraits and statements from their own personal networks into a visual map. The final exhibition was on display to the public in a dedicated gallery space for three weeks.

The project was rewarding for the artists, the museum and the group. It responded to and helped shape moves in the wider museum towards stronger collaboration and co-curation with audiences. As Peet articulates:

> It was a brave concept that was supported and executed brilliantly. It was particularly successful because all the artists realised the value of creating bespoke digital tools which enabled the young people to get high quality results quickly and easily, and which allowed them freedom of expression. Being able to see the high quality of the end result made participation exciting and more fulfilling.

Other success markers included a diverse range of activities, time to allow confidence and momentum to build, and trusting the group as adults with the technological equipment. The museum's and the artists' commitment to the group was essential. It was something they valued and were keen to respond to (Muse Research, 2008).

- 'At school you just follow the curriculum, but here we have got to play around with so many different things and it opens you up to such creative ways of making art, and such a range of things too'.
- 'We are part of something and it makes me proud'.[8]

Conclusion: interruption feeds evolution

Museums develop by challenging their own information structures and hierarchies. Working with artists, as well as researching and responding to visitors' responses, affects this change.

We are working within a culture that is challenging historical divisions, distributions of power and assumptions. I perceive a key influence on interdisciplinary practice in museums to be connected to the growing influence of digital communications technologies including the World Wide Web. These media's convergent, systemic and networked nature – along with their mass accessibility – have contributed to a heightened cultural awareness of the interconnected nature of things and ideas. This has thrown the gauntlet down to museums or any knowledge 'service-providers' in the business of reflecting and affecting contemporary culture to be wary of offering blinkered single-point perspectives or positioning the one-to-many broadcast as superior to many-to-many.

I likewise firmly believe that the strategies of community, media and participatory arts bring significant influence to the remodelling of a museums culture that acknowledges engagement works best when people have the option to do things as well as see things and to find out for themselves. It would be great to see further research in relation to the 'cultural' and 'science capital' visitors bring with them to specifically interdisciplinary projects.[9]

If I have any conclusion, it is that merging contexts is an emerging field and that interdisciplinary projects require collaboration. There is a natural give-and-take required when bringing art into non-art contexts which demands an acceptance of a certain muddying of intention and meaning, or an embrace of inevitable 'cross-contamination'. These projects can expand everybody's view, but they can also frustrate, obfuscate or alter original intentions. And they don't always work. The pitfall of interdisciplinary practice is that even subject matter experts go back to square one once a new subject beyond their specialism is introduced. Artists and art specialists can exoticise the processes and (visual and written) languages of science, reducing its potency and making limp points. Science specialists can reduce artworks to their pictorial representational elements without reading the wider meanings, or allowing these to emerge through placement and interpretation. Experimentation and risk-taking are critical.

Within SMAP we aimed to create situations and propositions that would foster a culture of exchange and reciprocity between art and science, for large general audiences to consider, albeit within the microcosm of our specialised museum.[10] We hoped the artworks would either act as provocative elements, encouraging visitors to add their own questions to those of the artists, or as catalysts which, seen in our context, might offer unexpected entry points for visitors to explore science, as well as the wider meanings in the art. Many of our aims were successful, the art certainly offered new entry points and added to a richer and more stimulating environment, others arguably less so. Some evaluations note that unaided few noticed the art interventions amongst all the other weird and wonderful things and high concept design of the permanent galleries. But it is important to note that there have been very few evaluations specifically designed to measure the art on its own terms, and none of these measured art in non-art galleries so much impact remains unmeasured. Our learning has huge relevance to the expanded field of interdisciplinary practice in recognising that while aims may be shared, the languages practitioners, organisations and visitors are more comfortable using are often different. Time, resources and a willingness to take risks are important in understanding and attempting to close or similarly satisfactorily resolve any gaps. In 2030 I would dearly love to meet the young woman I'm sure is out there who visited the *Who am*

I? gallery case in 2010 about brain science and held planted in her mind the thought that human consciousness may be explored through scientific instruments and artistic investigations through photography, pen and ink or pencil. I would love to be able to map the influence that may have had on her.

Notes

1 The SMAP curatorial team 1999–2014 also included variously Ruth Fenton, Rhona Garvin, Sarah Harvey, Alice Naylor, Deborah Smith, Fiona Parry and Iain Pate. In 2017 Katy Barrett was appointed Curator of Art Collections.
2 The senior management team were Gillian Thomas, John Durant, Graham Farmelo and Heather Mayfield.
3 'Visitors have expressed desire for innovative multidisciplinary and cross media (sound and vision) offers' (TWResearch, 2014).
4 See also Morris *et al.* (2009).
5 See my previous paper, Redler (2009).
6 Pioneered at Science Museum (2011) by Gammon, B. and Patten, D.
7 See Morris *et al.* (2009).
8 Participants' comments, *Being Connected.*
9 In 2008 Tate led a survey on the educational resources on *Rothko* in 2008 which suggests different levels of experience, confidence or 'cultural capital' make a significant difference to readings of and value gained from texts (Scott & Meijer, 2009). See also Science Museum, King's College and BP research into the concept of science capital (Science Museum Group, 2016).
10 The SMAP website is accessible from National Archives' archive of UK government websites at http://webarchive.nationalarchives.gov.uk/20150203151538/www.sciencemuseum.org.uk/smap.aspx (Accessed 07 February 2018).

Bibliography

Cumberlidge, C. (1996) *Arts Policy*, London: Science Museum. Unpublished report.

Davis, S. (2009) *Art Labels in Who Am I?* London: Science Museum. Unpublished report.

Fisher, S. & Paynter, F. (2011) *Who Am I? Summative Evaluation 2010 Redevelopment*, London: Science Museum. Unpublished report.

Harvey, S. (2012) *Hexen 2.0 Exit Interviews* [interviews], London: Science Museum.

Jardin, B. (2014) Made Real: Artifice and Accuracy in Nineteenth-Century Scientific Illustration, *Science Museum Group Journal* [Online], Autumn, 2. Available at: http://journal.sciencemuseum.ac.uk/browse/issue-02/made-real/ (Accessed 10 July 2018).

Morris, Hargreaves & McIntyre. (2009) *Visitor Responses to Statuephilia . . . at the British Museum* [Online]. Available at: www.britishmuseum.org/pdf/Statuephilia_Summative.pdf (Accessed 27 January 2018).

Muse Research. (2008) *Evaluating Being Connected*, London: Science Museum. Unpublished report.

Redler, H. (2009) From Interventions to Interactions: Science Museum Arts Projects History and the Challenges of Interpreting Art in the Science Museum, *Jcom*, 8(2), C04.

Science Museum Group. (2016) *Our Approach and Science Capital* [Online]. Available at: https://group.sciencemuseum.org.uk/our-work/our-approach-and-science-capital/ (Accessed 13 April 2018).
Scott, M. & Meijer, R. (2009) *Tate Papers* [Online], 11. Available at: www.tate.org.uk/research/publications/tate-papers/11/tools-to-understand-an-evaluation-of-the-interpretation-material-used-in-tate-moderns-rothko-exhibition (Accessed 07 February 2018).
TWResearch. (2014) *Media Space Summative Evaluation*, London: Science Museum. Unpublished report.
White, T. (2013) *Shackleton's Man Goes South*, London: The Science Museum. Available at: https://blog.sciencemuseum.org.uk/wp-content/uploads/2013/05/ShackletonsManGoesSouth.pdf (Accessed 05 April 2019).

4.5 Case studies

4.5.1 Digital art

Media and content

Since the 1970s there has been a growing interest in Digital Art, i.e. art that makes use of digital technologies at various levels. Yet, since the very beginning, digital technologies became not only the media but also the object of this specific artistic research and practice.

Digital artists of the past and of today respond actively to the need of society to think critically about our relationship with the digital sphere. These artists focus both on digital objects/products and on the social processes that control and are controlled by them.

We could say that today we are living with and inside a vast, pervasive and diffused super computer that connects everything from satellites in space to mobile phones in our pockets. "The digital" shapes our culture at all levels, and science engagement institutions often look for artists that can convey this critical perspective through the very object of their discourse.

Moreover, the level of technological literacy in society is inevitably destined to increase, and artists and cultural institutions have the ability and responsibility not only to incorporate such technologies in their exhibition practices but also to frame a critical discussion on how we influence and are influenced by digital reality itself.

Case study:
ArtScience Museum at Marina Bay Sands, Singapore (Singapore)

Honor Harger, Executive Director

ArtScience Museum at Marina Bay Sands is a cultural landmark in Singapore. Its mission is to explore the intersection where art, science, culture and technology come together. Its striking lotus-inspired building, designed by Moshe Safdie, features 21 gallery spaces spanning 6000 m^2.

Future World: Where Art meets Science

A permanent exhibition which comprises 19 digital art installations that come to life through cutting-edge science and technology. The immersive and interactive exhibition allows audiences to become an intrinsic part of the artworks on display, as the installations dynamically evolve through their presence and participation. The transient and transformational environment of the exhibition reflects a rapidly changing world.

Description

In ArtScience Museum's permanent exhibition *Future World: Where Art Meets Science*, the creative threads by which art, science, technology and culture are inextricably bound are expressed in 19 art installations and interactive projects for children, created by Japanese collective, teamLab. TeamLab is a collective known for their innovative use of digital technology as a medium for art. They are made up of over 400 artists, engineers, animators, mathematicians and architects. Working as one, they constantly break new artistic ground, creating worlds of magic and metaphor.

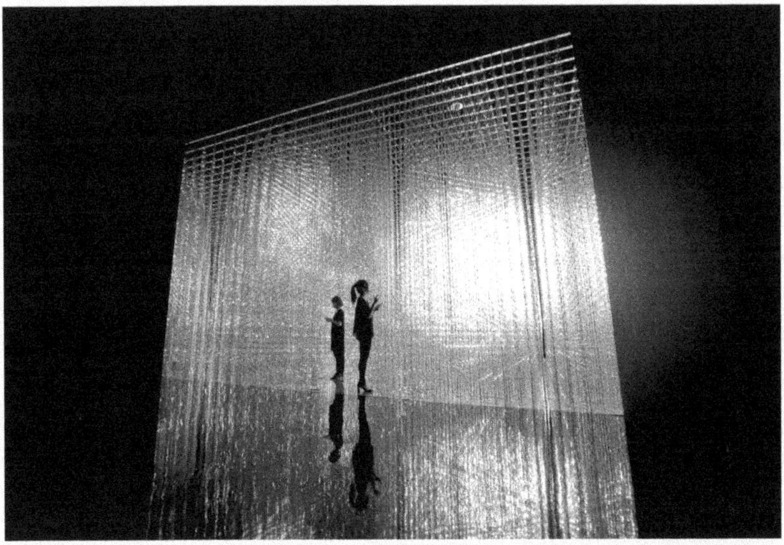

Figure 4.5.1 Crystal Universe, an installation consisting of an accumulation of lights, forming a sculpture that expresses the universe with which visitors can interact using their smartphones

Source: teamLab

Future World is a large-scale exhibition spanning 1500 m^2, organised into five zones: Nature, Town, Sanctuary, Park and Space, each of which uses cutting-edge science and technology to create constantly evolving environments that visitors of all ages are invited to enter and explore. As the artworks change with each interaction, each visit to *Future World* is unique. The exhibition invites visitors to have fun and to play, but also to reflect on their own position in relation to the natural world, other people, and the universe.

Future World was collaboratively conceived by ArtScience Museum and teamLab, and it took two years of integrated planning. The museum's team of around 50 people led the curation of the exhibition, the development of the interpretative texts, the design and delivery of the educational resources, visitor services, security, and all operational logistics. TeamLab's artists and technologists led the creation of all the artworks in the exhibition. Both teams worked together on the strategic positioning of the exhibition, the marketing and PR of the show, and the design of the physical environment.

Outcomes

The exhibition has become a symbol of the mission of ArtScience Museum as a place where art and science meet, and as a space to reflect on what the future might be shaping up. Together with teamLab, the exhibition was built as an environment that would change and transform over time, responding to each and every visitor, and drawing people to visit again and again. It was interesting to explore how an exhibition could be permanent, yet changing, created not with objects, but rather with digital technology and the participation of the visitors, who ultimately co-create each artwork with their own presence. Over time it is important to understand how to tend this sort of interactive environment, how to evolve and iterate existing works, and how to use education and programmes to broaden the stories told by the exhibition.

We need projects that create meaningful connections between people and their natural environment. Future World is intended as a place where those connections can be made; a place where art, science and technology meet; a realisation of Leonardo da Vinci's famous maxim, 'everything connects'.

If, as science engagement institutions, we are to begin to address the key systemic challenges that the world faces today – be it climate change or inequality – we necessarily need both the understanding generated by science and the emotional connections made possible through art.

4.5.2 Exhibit design and art

Science centres have a solid tradition in the development of so-called interactive exhibits. These exhibits encourage learners to act in various ways, both physically and mentally, to produce a change in the exhibit itself and/or in the understanding of a phenomenon.

Typically, interactive exhibits are connected to phenomena that pertain to the domain of physics, which can be directly and quickly reproduced and altered by users. Yet today there are also more and more exhibits that incorporate biological elements, digital elements and other components.

Somewhat similarly, there are a number of artistic currents that have placed the action of the viewers at their core. From Futurism and Dadaism to certain strands of kinetic art and contemporary virtual reality experiences, their outputs are incomplete without the viewer's interaction.

The collaboration between exhibit designers and artists interested in active audiences has often brought to the creation of successful installations that combine the invitation to investigate a specific phenomenon with an aesthetic experience and/or a reflective involvement that helps connect scientific/technological investigation to social/philosophical/personal matters.

Case study:
Centrum Nauki Kopernik, Warsaw (Poland)

Dagmara Szawdyn, Head of Science and Art Events Laboratory

Copernicus Science Centre is the first and largest science centre in Poland, opened in 2010. Its mission is to inspire people to observe, experiment, ask questions and seek answers. The centre has more than 400 exhibits grouped in six permanent galleries and temporary exhibitions.

Art + Science Collection

A permanent collection of 17 interactive artworks spread around the Copernicus Science Centre and among "classical" interactive science exhibits. Experienced artists from all over the world were asked to take their cue from the natural sciences, discoveries, and scientific theories to design artistic exhibits in which art and science intermingle.

Description

In 2008 Copernicus Science Centre decided to invest in the display of a new art collection to be placed into the exhibition spaces. This meant that visitors would interact with more traditional science exhibits and at the same time see and interact with a number of works of art.

Some items in this collection were purposely made for Copernicus and others were copies from works shown in festivals and other institutions such as Ars Electronica, but all were selected by members of the Copernicus exhibition team, that purposely looked for artists with experience in inventing and building interactive works.

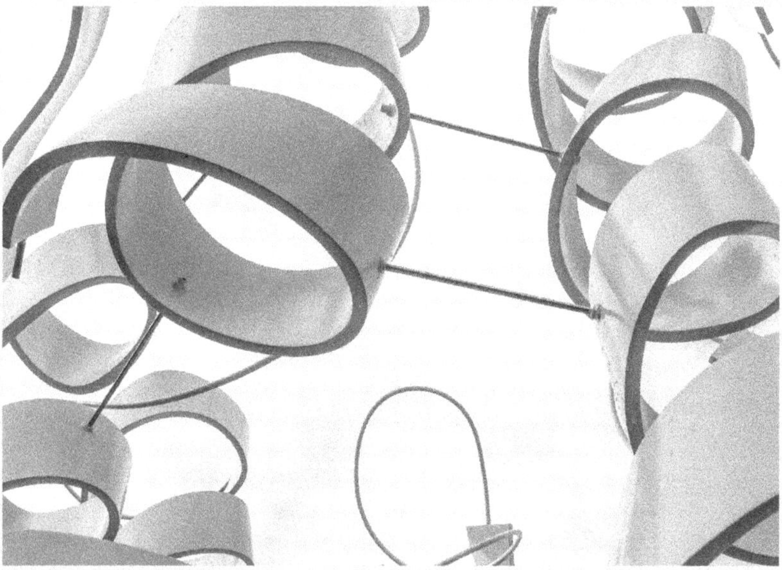

Figure 4.5.2 The Aquaporin, sculpture-fountain by Jaroslaw Kozakiewicz

Source: Photo by Agata Steifer/Copernicus Science Center

Artists were encouraged to seek inspiration within other sections of the science centre and then they were invited to design their work with the support of the Copernicus team.

The collection included: *Night Life*, Rufus Butler Seder (USA); *Freqtric Drums*, Tetsuaki Baba (Japan); *Voices of Fire*, Paul DeMarinis (USA); *Plant Engine*, Engineered Arts (UK); *Margot's Warsaw Cat*, Arthur Ganson (USA); *Light Waves*, Paul Friedlander (UK); *Wave*, Greyworld (UK); *Moths*, Piotr Jędrzejewski (Poland); *Eon*, Ksawery Kaliski (Poland); *Aquaporin*, Jaroslaw Kozakiewicz (Poland); *Copernichaos*, Mary Ziegler (USA); *A Neverending Story*, Marek Sułek (Poland); *Icy Bodies*, Shawn Lani (USA); *Reactable*, Reactable Systems (Spain); *Pipedream*, Bruce Shapiro (USA); *Leonardo*, Sharmanka (Russia/Scotland); and *Whisperers*, Elżbieta, Wela, Wierzbicka (Poland/France).

Outcomes

The display of the collection in the Science Centre has been a good occasion to observe how the public reacted to art works without being specifically aware of their presence.

With about 2500 visitors per day, every exhibit of the centre is constantly in use. The more delicate artworks did not survive this pressure, but the stronger pieces became part of the permanent galleries and are now used by science centre explainers to conduct specific workshops.

Successful exhibits were often the result of a collaboration between the artist, a scientist, but also the Copernicus team with knowledge of the exhibition conditions and of the audience itself.

The exhibited works seem to appeal to visitors that are sensitive to art, but they also help other visitors to discover a dimension of creativity and imagination.

In fact, one of the primary roles of works of art in a classic science centre context is to couple "hard science" with elements of enquiry about the future, social relations, ethical issues and so on. Art and science are like two sides of the same medal. Science may be the foundation of a science centre, through which it asks questions about the nature of the world, yet an artistic point of view can add relevant and innovative perspectives.

4.5.3 Re-imagining hard data

Metaphors are essential when we deal with information visualisation. If the object of our narrative is too large, small, complex or long-lasting it may become difficult or impossible for our senses to grasp. Thus science centre and museum curators often revert to metaphors that "translate" data into more sensory-accessible forms.

Many of the artists recently involved in residencies at science research facilities studying complex objects and data – such as elementary particles or outer space objects – have produced works that interpret information though sound, pictures, and moving images.

When data visualisation is used for scientific purposes or within the scientific community the correspondence between the resulting metaphor and original data must be accurate and unambiguous. In this case, in order for communication to be effective, there must be a strong correlation between the chosen visual metaphor and the viewer's own cognitive frames of reference.

On the other side, artistic metaphors may rely on looser connections to the viewers' preconceptions, making data more difficult to interpret without a supporting explanatory apparatus, but ultimately producing more emotionally intense responses.

Case study: FACT, Liverpool (UK)

Lesley Taker, Producer and Curator

FACT is a UK leading media arts centre, offering a unique programme of exhibitions, film and participant-led art projects. Showcasing groundbreaking new media art from across the world, its artistic programme has presented over 350 new media and digital artworks.

Unfold

An immersive and sensory installation which transports visitors into Space. Created by Japanese artist Ryoichi Kurokawa in collaboration with astrophysicist Vincent Minier at CEA Irfu, Paris-Saclay, the artwork is in fact an example of "experiential data visualisation" that helps connect complex science research to broader philosophical issues.

Description

Unfold explores the birth and evolution of stars offering viewers an artistic, yet scientific, representation of how the solar system was born and how our galaxy might evolve. The work was co-commissioned between FACT, Stereolux and University of Salford Art Collection, with the support of CEA Irfu, Paris-Saclay, Arcadi, DICRéAM and the support of the Goethe-Institut London.

In 2016, Vincent Minier and colleague astrophysicists at the Research Institute into the Fundamental Laws of the Universe (CEA Irfu, Paris-Saclay) granted Kurokawa the unique opportunity to explore data taken from giant molecular clouds in space – data which may hold the secrets

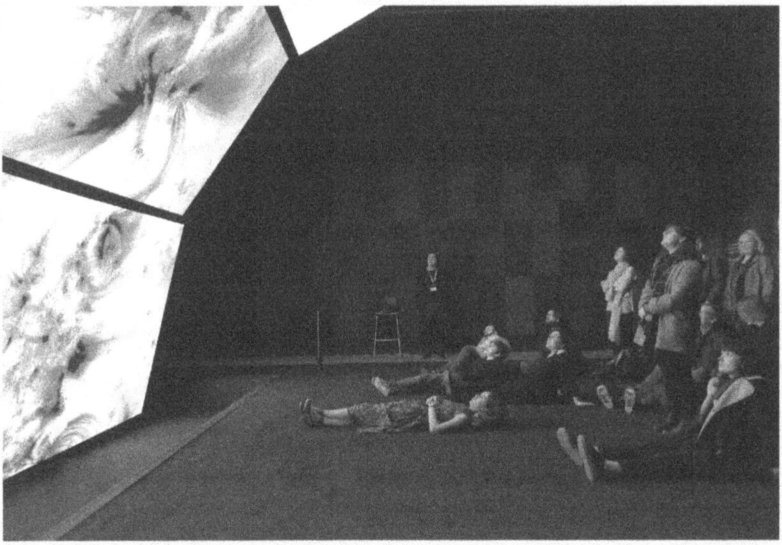

Figure 4.5.3 A frame of the immersive video and sound installation unfold, 2016, by Ryoichi Kurokawa at FACT, Liverpool

Source: Photo by Brian Slater

behind the birth of stars. Using these previously undocumented findings, Kurokawa constructed his vision of a molecular cloud as close as possible to the scientific truth using striking 3D representations of space, combined with interpretations of how star clusters form within the cloud until the birth of sun-like stars.

The resulting 8-minute-long film was presented as a large scale, sculptural multi-screen and surrounding sound work in a dark space created in FACT's Gallery 1. The artwork was associated to other iterations of the research (prints/audio work) which appeared throughout the FACT building and surrounding area. Visitors could lie on the floor to experience maximum bass and panoramic effect of the three-screen projection. This was 'feeling the universe' quite literally and created a truly embodied moment.

Outcomes

Scientists at CEA were already exploring the aesthetics of basic data visualisation for interdisciplinary science communication. Ryoichi's work with Vincent amplified the process of bringing aesthetics into scientific life.

We wanted to send out a message to the general public that art and science can maintain a strong philosophical meaning and speak to universal concepts and questions.

The experience created wonder and questioning in a broad audience, reaching local communities in Liverpool including science enthusiasts and academia.

Some of the creative scientists we work with recognise this value and hence are drawn to this type of activity. The work toured to the Jikji Festival in South Korea, Arts Santa Monica, Barcelona and other venues, signposting the commitment at FACT to working across science and philosophical subjects.

This exhibition presented a unique dialogue between science and art that allowed audience members to learn about cosmic phenomena and the formation of the solar system. It offered a platform for scientists and astrophysicists to open up their research to the public and raise awareness about their years of intense work, whilst pioneering multi-disciplinary research to create a ground-breaking and unforgettable audiovisual artwork.

Science and art might be treated as mutual starting points to move beyond 'traditional' data visualisation. Art turns linear science communication into an experiential practice. Artworks seem to leave those who have less experience of science theory with a stronger curiosity to know more. Though of course, embodied experience is, in itself, knowledge.

In this frame, art has the power to speak of complex theories and make them instantly recognisable: it brings into a person's previous boundaries a new field of experience and knowledge.

4.5.4 When the media influences the message

A popular exhibit in science centres – also known as "science on a sphere" – consists of a large globe on which data, static, and moving images can be projected.

The spherical globe can show real images, conveying data through a direct visualisation of reality. This is the case when projections include (sometimes live) satellites images of clouds, storms, night-time illumination or similar situations. In this case, data visualisation relies on the power and clarity of the real image itself. In other cases, the sphere can be used to project elaborated information connected to the geography of the globe, linking the visualised data to an immediately recognisable geographical distribution. In yet other situations, the globe can become a mere surface for the projection of other types of pictures that do not take into consideration the geographical factor, but simply make use of the spherical shell.

The flexibility of this tool allows for different types of descriptions, yet the medium somehow encourages certain types of stories, privileging for example geographical-related data organisation. This influence of the media on the chosen curatorial narratives is true of every technological media adopted. Mastering the tool creatively means intentionally letting the media inform our messages and stories.

Case study: Miraikan – The National Museum of Emerging Science and Innovation, Tokyo (Japan)

Yumi Miyahara, Project manager

Miraikan is a large science centre that aims to help visitors understand the things that are happening in our world today from a scientific point of view, pondering and discussing the future roles of science and technology through programs, talks, exhibitions and events.

Picture Happiness on Earth

An international 13-country project inviting students to collaborate, think scientifically about happiness and create multimedia representations for Miraikan's LED globe-like display. The program goal was to nurture a global mind, logical thinking and creativity, using aesthetics as a tool for data visualisation, communication and social activism.

Description

Between 2015 and 2017, the Project engaged 1509 students from 13 Asia-Pacific regions in analysing data to scientifically recount happiness in their countries. Online, 153,957 people voted to pick the best scenarios from this exercise and the winning creations were then programmed as video works by 188 female teenagers to be displayed on Miraikan's Geo-Cosmos exhibit (a spherical LED display 6 metres in diameter).

The programme consisted of three phases. First, educators from partner science museums came to Miraikan for a three-day workshop to learn about Miraikan's data, interactive tools, workshop content, and vision. At a later stage, 1321 students attended workshops to define happiness and create scenarios using science and statistical data. The best scenarios were selected through online vote.

To complete the project, 188 female teenagers in Japan participated in workshops to gain programming, graphic designing and video making

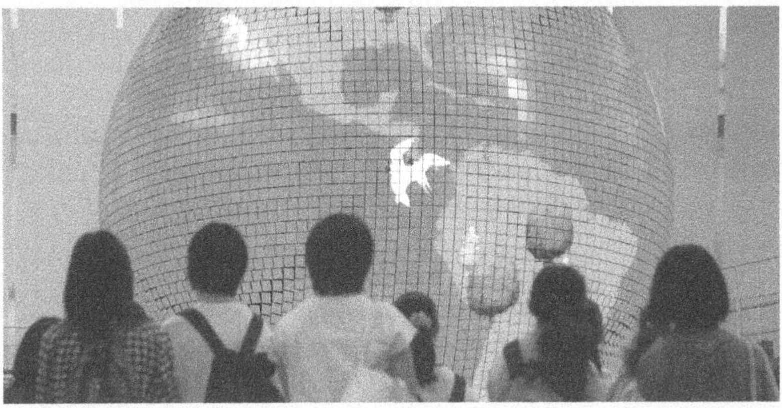

Figure 4.5.4 Scenery of video production

Source: @Miraikan

skills to create multimedia representations for Miraikan's symbol exhibit *Geo-Cosmos.*

Partner museums included Scitech (Australia), China Science and Technology Museum (China), Shizuoka Science Museum Rukuru (Japan), Petrosains (Malaysia), The Mind Museum (the Philippines), Science Centre Singapore (Singapore), Questacon (Australia), Chiba City Museum of Science (Japan), Busan National Science Museum (Korea), Science Alive! (New Zealand), National Taiwan Science Education Center (Taiwan) and National Science Museum (Thailand).

Outcomes

In order to communicate efficiently, the project worked using the expressive power of art to create visual works that would appeal to the audience's sensitivity. Through the process of conceptual creation, planning and directing of their work, teenagers received immediate guidance from a team of artists and creators on how to make their "sphere" videos attractive.

A variety of professionals worked together to make this project a success. The programme's basic idea formed in a brainstorm session with Bloomberg L.P., that later funded the project. Educators from partner museums and teachers from local schools recruited and facilitated workshops across the Asia-Pacific area while science communicators were in charge of helping students analyse data to make the scenarios empirical and able to connect with all participants. Curators gave lectures on how to create concise and powerful video works that could appeal to various audiences. Visual artists guided the whole process of making videos by using a script matrix, while graphic designers gave tips on design elements such as composition, drawing, colour scheme, and using editing software. Programmers gave guidance as to how to visualise the statistical data and transition graphical materials from a flat surface to a spherical surface. Corporate volunteers also helped with translation and programming.

The viewpoint of how to effectively appeal to an audience, ideas, and expression by professional artists greatly influenced not only the students involved, but also museum curators and staff. The project also offered an opportunity for teenage women to engage in STEM education, access science and develop both aesthetical and technical skills.

Global issues identified by youth – such as the U.N.'s Sustainable Development Goals (SDGs) – can be discussed, explored and investigated through artistic expression. Future generations need to be aware of what is happening in the world, think of every issue as a personal issue, discuss it and – last but not least – share it. In this perspective an artistic training can support the development of successful communication skills and encourage active engagement.

4.5.5 Humans and computational creativity

Parametric design is based on the study of how our influencing of a variable in a complex system changes the system itself. Architect Antonio Gaudi mechanically applied this method to his analytic study for the architectural design of the Church of Colònia Güell and many apply this idea in modern computing and software design.

Numerous artists focus on the creative process rather than on its outcomes. With specific reference to the technique of computer programming, such artistic research has worked on the production of "software", i.e. of tools, situations, rules that give other users the possibility to independently experience formalised processes that lead to unique results.

When thinking about computational creativity we can reason on the role of the human designer in this process, ultimately asking ourselves who the artist is. Is it the computer? The software designer? The software user who changes the parameters and ultimately creates a unique product and output?

Computational creativity works at the border between human and machine, investigating both the algorithmic nature of human creativity and the creative potential or influence of algorithms themselves.

Case study: MOMATH – National Museum of Mathematics, New York City, NY (USA)

Cindy Lawrence, Executive Director and CEO

MoMath is North America's only cultural institution entirely devoted to the wonders of mathematics. Its exhibits, gallery and programs are designed to stimulate enquiry, spark curiosity and reveal the wonders of mathematics, illuminating the patterns and structures around us.

The Mathenaeum

An interactive exhibit which allows visitors to create aesthetically interesting products while engaging with mathematical terms and concepts. In this case, aesthetics and creativity are used as a means to put visitors at ease and help them approach mathematics by engaging them in a subjective sensory experience, ultimately aiming to inspire positive attitudes towards the topic.

Description

The Mathenaeum is an exhibit at the National Museum of Mathematics that presents math in a visual, artistic and creative manner, allowing visitors of all ages to apply a pallet of operations to transform basic mathematical shapes into unique three-dimensional digital sculptures. Anyone, no matter what age or mathematical background, can interact with this exhibit to create personal "works of art", some of which are then printed with

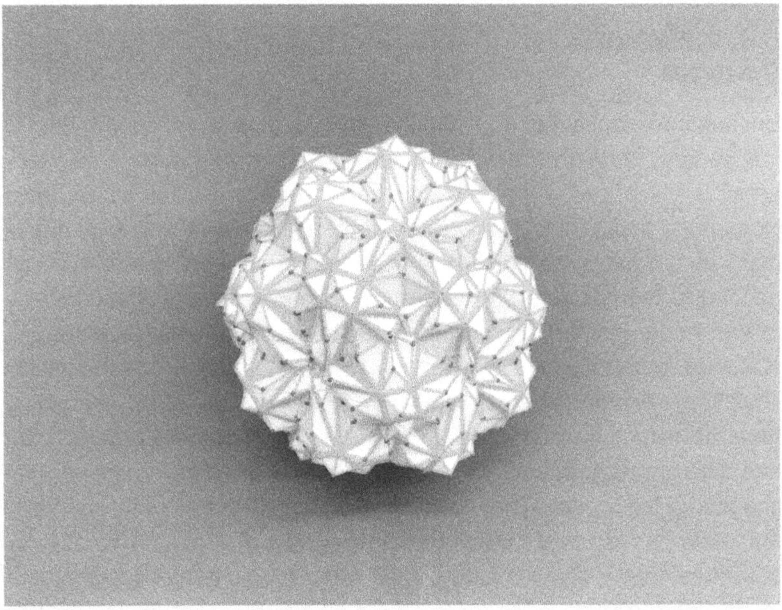

Figure 4.5.5 Mathenaeum exhibit, at the National Museum of Mathematics in New York City

Source: National Museum of Mathematics

the museum's colour 3D printer. The exhibit was installed in the museum when it opened in 2012. Electronic versions of the designs are currently being implemented to allow visitors to take home virtual versions of their unique creations. Visitors can also explore their work through virtual reality, stepping into their creations and exploring them from the inside.

Hundreds of thousands of visitors have had the opportunity to interact with the Mathenaeum, and small printed versions of visitor creations fill two showcases in the museum. Larger versions have served as auction items, trophies, and centerpieces for formal galas and events.

The Mathenaeum was designed by a team including Glen Whitney, George Hart, Tim Nissen and Cindy Lawrence. Glen and George provided the mathematical expertise, Tim designed the physical exhibit and Cindy worked on the user interface. Partner organisations included Lexington, who was responsible for the physical fabrication of the exhibit, and Three Byte, which handled the programming side. This exhibit was quite challenging to design from the technological point of view; consultation with museum trustees including mathematician Jade Vinson was also instrumental in creating a successful experience.

Outcomes

Because of its shape and making, the Mathenaeum contributes to the general MoMath mission of changing public perceptions of mathematics. Visitors are surprised to have an experience that allows them to express their artistry in a mathematical context and to experience math as a creative endeavour. There are no right and wrong answers, no long written explanations, and no cookbook recipes provided for visitors to follow. Instead, visitors are encouraged to be creative with form, structure, pattern and colour as they create their unique mathematical sculptures. Additionally, they learn new mathematical terms, everything from the names of various types of three-dimensional solids to the various mathematical operations that can be performed to manipulate the images on the screen. They are also encouraged to play with symmetry and geometry in a way that is novel, intriguing and engaging. Finally, the opportunity for visitors to vote on the day's user-created designs, with top sculptures being 3D printed and put on display in the museum, adds an element of excitement to the exhibit.

Many people think of math (and more in general of hard sciences) and art as polar opposites: math is dry, boring, rote and black and white, whereas art is interesting, colourful, creative and beautiful. Combining art and math in interactive and engaging exhibits helps break down these artificial barriers and overcome these stereotypes. Art allows us to re-interpret scientific processes as creative, non-linear and open-ended experiences.

4.5.6 Artistic activism and narratives on environment

There is no unique and comprehensive definition of Environmental Art, yet this research is often conducted by artists who are personally concerned with the conditions of the environment and who seek not only to explore the relationship between mankind and the natural environment, but also to actively contribute to change human attitudes towards nature.

After the first alarm bells for the evident loss of biodiversity went off in the 1980s, environmental conservation programs in the 1990s advocated for a deeper integration between the ecological and social dimensions, as well as for a more direct involvement of scientists in the fields of environmental decision making and education. An activist strand of artistic research then formed a generation of artists that have joined this process, engaging not only in their own practice but also in education.

The growing number of academic interdisciplinary courses feeds on this dedication and on emerging collaborations between scholars from the fields of visual art, literature, philosophy, sociology, botany, conservation and so forth. It is also through the training of future scientists that artists can become active members of social change.

Case study: Parque Etnobotánico Omora - Centro Universitario de la Universidad de Magallanes, Puerto Williams (Chile)

Ricardo Rozzi, Director of the Sub-Antarctic Biocultural Conservation Program, Universidad de Magallanes, North Texas University
Paola Vezzani, Art-Researcher for the Sub-Antarctic Biocultural Conservation Program, Universidad de Magallanes

The Biocultural Conservation Program based at the Parque Etnobotánico Omora aims to study and implement activities to generate conservation consciousness. It is led by an international and interdisciplinary cooperative group that integrates biological investigation, philosophical concepts and arts.

Ecotourism with a Hand-Lens

An outdoor ecology and ethics university course that translates into a public outdoor activity created by a team of artists, writers, scientists and philosophers to promote sustainable development, including wellbeing of both humans and the community of living beings. Drawing and metaphor composition are used both as tools for observing and understanding the world, thus reinforcing scientific knowledge and ethical valuation; tools for science communication and cultural transformation.

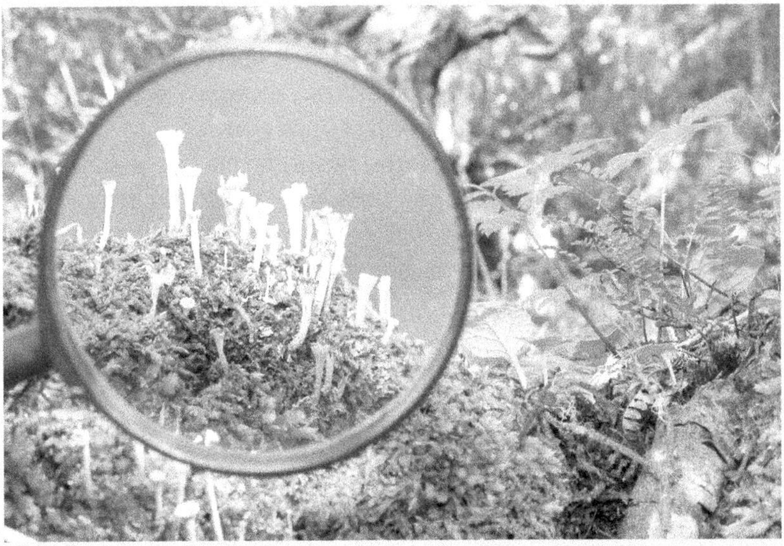

Figure 4.5.6 Ecotourism with a Hand-Lens: an innovative activity developed by the interdisciplinary team of artists, philosophers, and scientists at Omora Park. It invites visitors to observe the microcosm and appreciate the ecological, aesthetic, economic, and ethical values of biological and cultural diversity, and their interrelations

Source: Photo by Adam Wilson, Archive Omora Ethnobotanical Park

Description

Located in the Cape Horn Biosphere Reserve in Chile, the Omora Ethnobotanical Park is an outdoor museum and a biocultural laboratory dedicated to research, education and conservation at the southern end of the Americas. Every year since 2000, about 30 university students are offered the *Ecotourism with a Hand-Lens (EHL)* course. Teachings involve philosophers that discuss environmental ethics and scientists who develop research on mosses, birds, insects and the methodology of minimum alteration of the environment and of its inhabitants. In order to communicate the values discovered during their field experience, students belonging to departments of arts, philosophy, journalism, political and biological sciences work with a writer to develop metaphors and then with a visual artist to transform them into images and prepare interpretive signs that are later installed in the park itself. This experience is shared with local school children, teachers and ecotourism guides who jointly guide tours based on the work done by students and conducted by artists, philosophers, scientists, members of the Yahgan indigenous community and by the students themselves. These courses are offered year-round and help appreciate the aesthetic, economic, ecological, and ethical values of the Miniature Forests of Cape Horn.

EHL courses and activities both rely heavily on the use of drawing as an observation and discovery tool. Metaphorically, the course's lenses do not only amplify the vision of mosses and other small organisms, but also offer conceptual lenses to broaden the understanding of nature and ethical co-inhabitation within it.

Outcomes

Bringing writers and visual artists to teach EHL courses alongside philosophers and scientists allows students to participate in truly transformative experiences. As they explore the power of drawing, narratives and metaphors, they enrich their biocultural understanding while also contributing to conserving biocultural diversity itself. Science and philosophy teach them that mosses, humans and all living beings share the common vital pulse of cellular respiration, growth and reproduction. Drawing and composition of metaphors help them to perceive the breathing of mosses, the calls of birds, the waves of oceans and many contrasting human languages.

Complemented with philosophical readings and scientific enquiry, the EHL courses and resulting guided activities help participants experience and understand first-hand different perspectives through which to perceive and value the world.

The way the course is structured and feeds into the public activity of the park results in a win-win situation as students feel integrated and responsible for the environment they have analysed and worked in, while the park finds new creative and often unexpected ways of communicating to local ecotourism guides and the public its values of biocultural diversity.

Drawing is a medium available to everyone, and in the intimate time taken to observe and prepare a drawing, the artistic practice becomes a tool of personal observation to recognise otherness. Emotions stirred by such artistic practices can enhance future science researches' skills to understand the surrounding world and respect other living beings as co-inhabitants rather than as mere natural resources.

4.5.7 Theatre and teaching

In the 1950s British teacher Brian Way founded the Theatre Centre, a company made of professional actors that would perform for school children with the specific purpose of being educational and integrating the school curricula where possible.

The idea behind Way's work was that performances could be entertaining and informative at the same time, triggering emotions as well as conveying information. Often the students would be addressed directly by the actors and then be invited to participate through reflection and debate during the play or in follow-up activities.

Since then, performances targeting school audiences have become increasingly popular and have diversified in shape and form, sometime crossbreeding with other forms of traditional or popular art and giving birth to more specific local expressions such as stand-up comedy, dance performances, audio-visual plays, storytelling monologues, musicals and of course . . . "science shows" now popular in many science centre settings.

Case study:
Science Centre Singapore, Singapore (Singapore)

Tit Meng Lim, CEO

Science Centre Singapore is a non-formal educational institution and leading regional science centre. Since the 1970s it presents the wonders of science, technology, engineering and mathematics through a blend of exhibitions, educational programmes and events.

Sex Cells – A Science Musical

A science musical based on the biology of human reproduction. The show aims to demystify how babies come about through the union of a sperm cell and an egg cell. In the play script, the journey of the sperm successfully reaching and fertilising an egg is met with numerous challenges and hazards.

Description

The show begins with a young girl and her father in a maternity nursery filled with many expectant mothers and new-born babies. The opening song and dance is about the joy of a new-born. Looking around, the girl suddenly asks her father where babies come from. Feeling a bit embarrassed but not wanting to hide the facts on how babies are made, he explains to the girl how his own parents produced him. The opening scene then transits from the father's narration of his parents' wedding to their sleeping together in the bridal chamber when the theatre suddenly darkens and the scene bursts into the cellular events inside the reproductive system. Actors playing the sperm cell characters rush out as soldiers with a mission to search for their target egg. The sperms have to compete with each other along the way. A big ball representing the egg is rolled down from the audience seats with the audience pushing

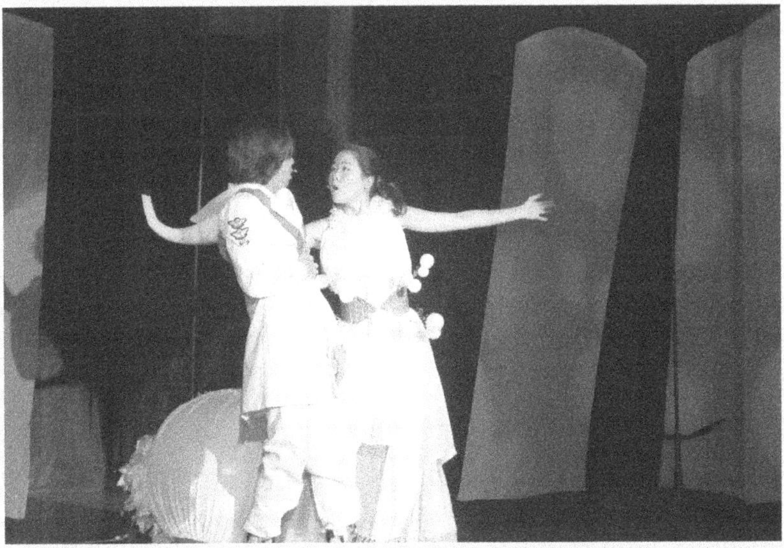

Figure 4.5.7 A duet scene where the champion "sperm cell" finally meets the "egg cell"

Source: Science Centre Singapore

the egg onto the stage to simulate ovulation. The egg character sings a song, wondering what will happen to her while she waits patiently. When the champion sperm finally meets the egg, a duet song and dance depict the fertilisation process. The successful union scene then transits back to the dialogue between the girl and her father, telling how everyone is born with unique genetic combinations from different eggs and sperms.

Ten members of the science centre staff were selected to act alongside the professional cast. The script was written by the science centre CEO, a biology professor, derived from his university lecture titled 'Sex at the cellular level'. He was supported by science educators as producers and venue managers.

Outcomes

The musical created curiosity and interest about biology and moral ethics. It fitted the science centre's mission to promote interest in science. The musical project was devised firstly to get the staff to think creatively beyond the typical science centre show. The objective was to make use of a powerful edutainment platform, especially fusing science and performing arts, to reach out to a general public to make science accessible to many.

In-house competences in storytelling, showmanship and communication skills were raised and the staff involved learned a lot from the professional artists. Practises and rehearsals were conducted after office hours into the late nights. The demand for commitment was very high but the outcome and reward made all the sacrifices and hard work worthwhile.

The project built confidence in the team and promoted cross departmental collaboration. The show had even the finance director, managers and administrative assistants performing alongside science educators and professional actors. It was a very enriching experience for all the staff.

The experience also created new partnerships with communities outside the science centre and opened the mind of people who had biased feelings about science and science centres. The show managed to engage audiences across social strata and educational background as the topic on sex and reproduction is a fundamental and universal theme that all are intrigued and curious about. In fact, school students should receive sex education, but many teachers find it hard to bring out the topic on sexual reproduction. Some parents also find the topic slightly challenging to explain to their children clearly.

The use of a musical as a means to talk about this topic, allowed to create a powerful impression that a typical textbook narrative or science exhibit on sexual reproduction could not have created. The show also delivered an ethical message on an emotional level, promoting respect for life and understanding of how everyone is genetically unique. The emotional power of artistic representations should be kept into consideration and treated with care when dealing with scientific topics and the effort to promote critical rational reactions. Follow-up work, reflection and dialogue become key means to balance emotional reactions with debate and facts, especially when dealing with ethical-related points of view.

4.5.8 The power of contemporary art gallery design

Contemporary art museums have often underlined the value of artworks by placing them in specific settings, illuminating them, isolating them, giving them architectural emphasis, just as one would do with an altar or an object of awe. As soon as one experiences the object in such an environment, one instinctively recognises the underlying value-message connected to the item. In contrast, in traditional natural history and science museums, precise and accurate scientific explanations are frequently crafted to instruct about the object(s) at hand, commonly displayed in rather didactic-driven ways. Is it possible to have these two approaches overlap in a museum where art and science meet to engage its visitors with the objects, concepts and phenomena being addressed, based on an immediate appreciation of their value and meaning?

Jorge Wagensberg and others have advocated for the application of a design – typical of contemporary art galleries – in science exhibitions. With this technique, the value of scientific items – such as natural specimens, technological devices, or even concepts – is communicated, from the start, through the exhibition setting and design itself.

Case study: The Hall of Biodiversity, Porto (Portugal)

Nuno Ferrand de Almeida, Director

The Hall of Biodiversity, located in the Botanical Garden of Porto, is part of the Natural History and Science Museum, University of Porto, and is also integrated in the national network of Ciência Viva Centers. The Hall of Biodiversity invites everybody to celebrate life.

Telling Stories about Life

A museographical and museological project for an entirely renovated natural history museum building that integrates elements of design mutated from the artworld to reinforce immediate emotional connections with abstract and ethics-related contents.

Description

The Hall of Biodiversity represents a journey through life. The key idea of its permanent exhibition – the first part of the new Natural History and Science Museum – is to explain the importance of preserving biological diversity on our planet.

By touring the Hall of Biodiversity, visitors discover the countless shapes, sizes and colours that characterise organisms today. They can also gain an understanding of what is natural, artificial or sexual selection, what is a species and how it is formed, what is domestication and the transformation of life, and how humans have evolved.

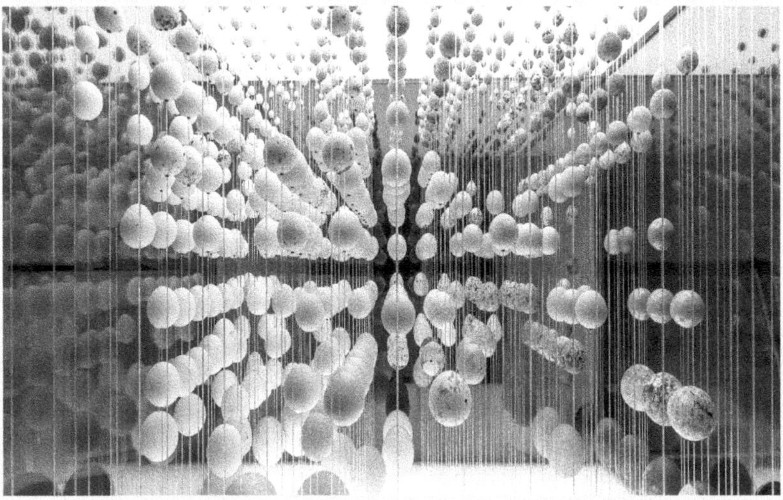

Figure 4.5.8 Why should we preserve biodiversity? At the Hall of Biodiversity, a dazzling display filled with real eggs of all shapes, colours and sizes, brings together art and science to unveil how the answer to this question could simply be because of the stunning beauty life entails.

Source: Photo by João Ferrand

The Hall of Biodiversity was inaugurated in 2017. The restoration of the 19th-century palace was coordinated by a renowned architect, while the museography project resulted from the work of an extended and multidisciplinary team directed by a researcher, a museographer and an artist. In addition, a landscape architect worked on the dialogue between the historical palace and the gardens that surround it. Today, science explainers and communicators, guides, researchers, other museum professionals, and artists, all participate and interact with the public by providing experiences and enacting narratives about life in the attempt to fascinate and enlighten visitors of all ages and backgrounds.

Outcomes

The main message we wanted to convey in The Hall of Biodiversity referred to the importance of the conservation of biological diversity on our planet. To do so, in the frame of a new Natural History and Science Museum, a different approach was needed and art seemed like the obvious choice. Today, every single story about life is presented to visitors after they have been conquered and fascinated by a captivating art installation.

The Hall of Biodiversity has set the way for the building of the new Natural History and Science Museum, University of Porto and, while this new global institution will have major sections dealing with collections, explorations, education and exhibitions, a permanent multidisciplinary team of researchers, artists and science communicators will work together across those sections to find new ways to interact with the public and to ensure a greater impact and a wider scope of the messages we want to convey.

Using museographical standards that are typical of art settings can help convey messages more directly and powerfully. Most viewers react to modern aesthetic standards (e.g. spatial rarefaction and minimalist forms) with a spontaneous and almost unconscious recognition of value so that abstract messages, such as the value of biodiversity, can be passed on immediately and through emotional responses that may precede more analytical and detailed investigations of the topic.

5 Art for science education and enquiry

Patterns of thinking

There is ample literature discussing the specific role played by art in general learning processes. For one thing, art is by nature prone to widen discourses and include criticisms, scepticism, uncertainty and so on. Moreover, artistic research has intrinsic characteristics that seem to resonate well with educational theories that encourage personal and creative enquiry. This section focuses on examples of situations in which museum visitors are somehow actively involved in the artistic process itself, either through an "encouraged" interpretation of the artwork, an interaction with an art-based exhibit, participation in an artistic research/performance or through the direct making of an imaginative product.

As Marina McDougall explains, the Exploratorium in San Francisco (USA) has developed an approach based on "art as enquiry" with the ultimate aim of empowering its audiences. The initial interest of the institution's founder, Frank Oppenheimer, in investigating art as both an object and a process led to the development of a multidisciplinary cultural setting and to specific research on how and why art-influenced interactive exhibits represent effective educational and pedagogical tools, which develop curiosity and understanding through direct experience and creative learning.

Based on the extensive work done by Ars Electronica in Linz (Austria), Maria Pfeifer and Christopher Lindinger illustrate four "Art Thinking Strategies" that may be adopted to develop critical thinking on scientific topics within museums that work to become "discursive venues". Their chapter focuses on the role played by art in these strategies and on its power to connect scientific and technological topics to reflections on the future of society, providing and stimulating what they refer to as an 'artistic view of the future [that] has especially high potential to formulate solutions to problems that have not even emerged yet'.

At the Qattan Foundation Science Studio in Ramallah (Palestine), art-science educational workshops are offered with the underlying notion that self-expression is a powerful tool to reinforce both learning and identity. In a complex socio-political setting, art-science activities offer participants the opportunity to focus on technical-scientific aspects while simultaneously allowing them to express their personal views. As Samar D. Kirresh points out, art helps bring into science learning an individual meaning-making attitude that is ultimately functional to the improvement of the learning process itself.

Ian Brunswick and Andrea Bandelli recount how Science Gallery Dublin (Ireland) – a landmark institution in the evolution of hybrid cultural contexts – is investigating the specific added value brought by visitors to art-science exhibitions. The authors discuss the process that leads to the choice of exhibition themes through audience observation and empathy. In this frame, the institution becomes a space where artists and scientists are invited to 'think of the gallery as a two-way dialogue that can contribute to their research, influence their practice, and be more than an instance of "selfless" public outreach'.

The case studies in this section present practical examples of techniques and experiences where art and science merge to create an emotional connection with a scientific topic, to reinforce learning through personal understanding or direct interaction, to give space to individual voices while facilitating cooperative work.

5.1 The Exploratorium

Art as inquiry

Marina McDougall

Introduction

The integrated arts learning methodology of the Exploratorium in San Francisco, California (USA), represents a strong, sustained model of inquiry-based learning through art and science. Situating the participant's perception and awareness at the core of each experience, the Exploratorium specialises in multidisciplinary approaches to learner-centric design.

A sense of play, experimentation, and freedom pervades both the Exploratorium's popular public environment, as well its behind-the-scenes professional development work with educators. While many elements contribute to the Exploratorium ethos, art plays a vital role and is fundamental to its learning philosophy. Artists have added to the institution's wide appeal, its sense of wonder, and the depth of engagement for its publics. As writer Dave Eggers described: 'The arts are the secret sauce of the Exploratorium'.[1]

The Exploratorium approach recognises the arts as a powerful vehicle for captivating attention and inspiring curiosity in the world around us. Art can be transformative. It can change our perception by opening up space for the imagination, for rumination, and new ways of seeing. Art can imbue experience with a sense of poetics, with sensorial richness, and intellectual insight; and can help us approach complex questions in creative ways.

The sensibilities, insights, and discoveries of both visiting and staff artists have influenced the Exploratorium's working methodologies, hybrid culture, and public experiences. Since its inception, artists have participated in nearly every dimension of its work. Artists-in-residence (AIR), working as part of interdisciplinary teams, have contributed to the conceptual growth of the Exploratorium's educational R&D work. Artists collaborating with staff have developed hundreds of projects, many of which have become incorporated in the Exploratorium exhibit collection. Artists have created large-scale public artworks for the Exploratorium outdoor, civic space (most recently through the Over the Water program); and developed hands-on activities (as artist collaborators working in the Tinkering Studio, for example). Artists regularly present new work in temporal public programs (such as the Cinema Arts, Resonance, and After Dark series), and have been featured in temporary exhibitions.

While in our midst, artists have recently investigated: theories of colour perception (Lucky Dragons); exchange value (Paul Ramirez Jonas); evolutionary biological processes (Theo Jansen); tide mechanics (Tim Hawkinson); eye floaters (Nina Katchadourian); carbon and air quality measurement (Rosten Woo); sea level rise (Sara Dean and Byron Au Yong); and micro-meteorites (Tristan Duke). Recent projects have taken the form of an immersive fog sculpture (Fujiko Nakaya), social benches (Jeppe Hein), a pedagogical walk (Harrell Fletcher), performative scores (Zarahouie Abdalian), a library of rocks (Ilana Halperin) and a live eclipse sonification (Kronos Quartet). These artworks frame subjects in new ways and capture public interest through a wide range of emotional modalities from the intellectually engaging to the humorously wry, to the poignantly existential, to the politically charged.

A look at the Exploratorium's arts philosophy, institutional formation and hybrid culture suggests that when art is genuinely valued as contributing to knowledge, it can play a vital role in public institutions dedicated to science and interdisciplinary learning.

Frank Oppenheimer and art at the Exploratorium's foundation

The Exploratorium's history has influenced our current working processes where artistic methods and aesthetic sensibilities are incorporated into the development of new areas of research and public offering. Visionary founder Frank Oppenheimer seeded the conditions for this interdisciplinary approach to thrive by cultivating a value for the arts at the Exploratorium's inception. Understanding Oppenheimer's personal background, the institution's roots in science education reform, and how the arts became incorporated into its pedagogical approach has been helpful to us in orienting the current work of the arts at the Exploratorium and envisioning its future.

The Exploratorium's distinctive culture took root in the late 1960s and early 1970s, a fertile time for utopian thinking, social experimentation, and new projects at the juncture of art and science. Oppenheimer, a physicist and educator steeped in the arts (Oppenheimer, 1976), drew upon the combined capacities of scientists, engineers, artists, designers, and teachers to shape a new kind of public museum based in experiential learning.

Before founding the Exploratorium,[2] Oppenheimer had already had several careers and lived U.S. history writ large. In 1945 he joined[3] the secret Manhattan Project at Los Alamos under the direction of his renowned older brother, J. Robert Oppenheimer. Together they lay side by side in a bunker at Trinity test site and witnessed the haunting first detonation of a nuclear weapon (Cole, 2009, pp. 59–60).

Later, Oppenheimer researched cosmic rays as a professor of particle physics at the University of Minnesota. In 1949 he resigned under pressure from the House Un-American Activities Committee due to his former association with the Communist Party (Cole, 2009, p. 9). Blacklisted during the McCarthy era, Oppenheimer's family took up cattle ranching in Pagosa Springs, Colorado. A decade later in 1957 he became a high school science teacher, and in 1959 after the political climate had cooled, was offered a position at the University of Colorado (Hein, 1990, pp. 11–12).

Oppenheimer's participation in a national postwar effort to remake science education during this time would later become an important influence on the future Exploratorium (Hein, 1990, pp. 12–14). In 1958 Oppenheimer became a chief collaborator to physicist Jerrold Zacharias at the Massachusetts Institute of Technology on the development of a project called Physical Science Study Curriculum[4] (1958–1968) to reform science learning in secondary education across the U.S. The project was followed by Elementary Science Study (1960–1973), which focused on transforming approaches to elementary school science curriculum. These nationally scaled projects emphasised an experiential approach to learning. The participation of artists such as photographer Berenice Abbott and filmmaker Richard Leacock lent the work a poetic dimension grounded in the wonder of everyday phenomena.[5] Through these projects[6] Oppenheimer garnered a sense of how artists could participate in the development of experientially based, pedagogical approaches.

In 1965, Oppenheimer received a Guggenheim Fellowship that allowed him to research historic science museums in Europe.[7] During his travels, a vision for a new kind of public learning laboratory began to emerge. Upon his return Oppenheimer convinced San Francisco civic leaders to allow him usage of the vast 90,000 square feet Palace of Fine Arts. The Exploratorium opened in 1969.[8] An informal wooden sign reading: 'Here is being created, a public museum dedicated to awareness', was hung in the entry way to the cavernous space.

Oppenheimer's friend and biographer K. C. Cole observed that everything that had ever happened to him in his life went into the Exploratorium.[9] Oppenheimer's remarkable experiences influenced the social vision behind the Exploratorium and the core values that it holds to this day – the belief that a curious, and confident public inspired to explore and understand the world around them, will actively participate in public conversations addressing the critical issues that we confront as a society. For Oppenheimer adopting the creative mindset of the scientist and artist – latent in each of us – is crucial to the project of imagining a better world.

Cole's biography *Something Incredibly Wonderful Happens: Frank Oppenheimer and His Astonishing Exploratorium* (Cole, 2009) – as well as his own impassioned writings – give insight into how Oppenheimer's personal experiences influenced his belief that the discoveries of artists were as important and valid as those of scientists; that art was a 'basic form of knowing and learning . . . an essential mode of communication and discovery' (Cole, 2009, p. 191). Oppenheimer described artists and scientists as 'the official "noticers" of society – those who help us pay attention to things we've either never learned to see or have learned to ignore' (Oppenheimer, 1977). He decried the undervaluation of art in U.S. public education, stating: 'If art education is to be considered basic in schools and museums, we will require more than new curricula and improved teacher training. We will require basic changes in societal values'.

Early arts program

Artists were important contributors to the Exploratorium from its very beginning. Reviewing the inception of the Exploratorium's arts program suggests that when the arts are foundational, they can become core to an institution's approach.

The Exploratorium's official opening in October 1969 was timed with the San Francisco debut of the exhibition *Cybernetic Serendipity* (Cybernetic Serendipity, n.d.). Curated by Jasia Reichardt for the Institute of Contemporary Art (ICA) in London in 1968.[10] The exhibition featured the explorations of artists and scientists alike – architects, mathematicians, engineers, poets and composers – intrigued by the creative possibilities of computation. *Cybernetic Serendipity* presented Oppenheimer with a concrete sense of how artistic investigations – those of Allison Knowles, John and James Whitney, Nam Jun Paik, and many others – could manifest as engaging, non-didactic, interactive experiences illuminating new areas of technological exploration for the public. Ideal for the Exploratorium's participatory approach, 'the observer became a constituent of the artwork' (Hein, 1990, p. 34). For example, the design of *Drawing Board*, developed by John Ravilious, anticipates public participation in the creation of Lissajous figures made by placing felt markers on a drawing board suspended by swinging pendulums. Originally arranged to be on view for six weeks, *Cybernetic Serendipity* remained at the Exploratorium for two years. At the end of its run, ten works from the exhibition were acquired and incorporated into the Exploratorium collection. Several works are still in the public exhibit environment today.[11]

Even in its first few months of operation the Exploratorium became known for its burgeoning art scene. Hein writes:

> Even before the opening, the Exploratorium had acquired a reputation as an alternative exhibit space and performance center, and a series of unconventional art events took place there during the first few months of the museum's existence. Artists were drawn to the experimentalism of the Exploratorium and to Oppenheimer's energy and inquisitive intelligence. Many artists were eager to collaborate with scientists and technicians and to learn from them, and a rapport between the museum and the San Francisco arts community developed rapidly.
>
> (Hein, 1990, p. 150)

Experiments in Art and Technology (EAT), formed in New York in 1966 by engineers Billy Klüver and Fred Waldhauer; and artists Robert Rauschenberg and Robert Whitman with a mission 'to develop collaboration between groups unrealistically developing in isolation' made contact with the Exploratorium in its first months of operation. EAT's West Coast branch led by John Almen and Merlin Stone assembled a show of a dozen works that opened together with *Cybernetic Serendipity* in the fall of 1969.[12]

In spring 1973 the Exploratorium presented the traveling exhibition *Multiple Interactions*. The project was organised by György Kepes, a visual designer who led another groundbreaking art/science program of the era, the Center for Advanced Visual Studies (CAVS) at Massachusetts Institute of Technology (MIT) (MIT's Center for Advanced Visual Studies, n.d). *Multiple Interactions* presented works developed by CAVS fellows – including Stan VanderBeek, Ken Knowlton, Otto Pienne, Alan Sonfist, and Wen-Ying Tsai. One of the exhibition's most popular works, *Flame Orchard*[13] was a sculpture constructed of aluminium tubes filled with propane gas. 'When the gas was ignited it formed a fiery record of wave shaped sound patterns' (Hein, 1990, p. 153). Supported by a grant from the National Endowment for the Arts (NEA),[14] *Multiple Interactions* was a rare art and science show presented at both art and science venues across the United States.

The Exploratorium was an ideal environment for presenting these hybrid art and technology projects of the late 1960's: each generated a productive dialogue among disciplines; evoked emerging spheres of exploration; and allowed for new ways of seeing. Hein describes how

> The works displayed in *Cybernetic Serendipity* were treated as art by the art critics, but they fell between art and technology, and neither their subject matter nor their mode of production clearly warranted their classification as one or the other . . . Their ambiguous status exactly fit the philosophy of the Exploratorium, which promoted that same ambiguity.
>
> (Hein, 1990, p. 37)

Artworks at the Exploratorium continue to occupy this intriguing nether zone leaving arts staff to weigh standard art world paradigms against pedagogical ones.

The technical, interactive works in all three exhibitions – *Cybernetic Serendipity*, the EAT show and *Multiple Interactions* – required maintenance beyond the capabilities of most conventional art venues which privileged the presentation, protection, and conservation of valuable cultural artefacts. By contrast, the Exploratorium, with its roots in the development of pedagogical tools for educational contexts, placed greater importance on public experience – emphasising the role of art as a medium for furthering curiosity and understanding. Since the success of the Exploratorium participatory environment depended upon the effective functioning of exhibits and their durability, the Exploratorium technical staff was adept at repairing and even improving upon the works presented in *Cybernetic Serendipity* (Hein, 1990, p. 153). Later, when artworks from the exhibition became incorporated into the Exploratorium collection, they were re-contextualised and sometimes even renamed.

By 1972 the Exploratorium began using the subtitle *a museum of science, art, and human perception*, and Oppenheimer began working with Peter Richards[15] to initiate the Exploratorium's Artist-in-Residence (AIR) program.

One of the first projects, *Sun Painting* (1971) was developed by self-described "natural philosopher" Robert Larue (Bob) Miller with help from Exploratorium engineers. The project turned upon the building of a photoelectric heliostat that would trace the arc of the sun over the course of the day (Hein, 1990, pp. 154–158). A hole was cut into the Exploratorium roof to beam the sunlight into the building, where it was reflected on plane mirrors, dispersed by prisms, and projected onto screens. During this period Miller, a charismatic teacher and physicist by training, began considering light as information and leading his spellbinding, metaphysical *Image Walk* (Exploratorium, 2018a). Miller's inventive props and prompts allowed participants to experience the everyday phenomena of sunlight and shadow in confounding ways. Miller's poetic sensibilities provided learners a direct, simple, non-didactic, way to explore light and perception (Hein, 1990, p. 155).

In the first years of the AIR program Oppenheimer and Richards also engaged artist and art education advocate Ruth Asawa to lead a series of public workshops exploring geometry through physical manipulation of form with everyday materials (Exploratorium, 2018b, 2018c). (Later in the 1980's School in the Exploratorium, a professional development program for educators,[16] worked with artists Mildred Howard and Ann Chamberlain to incorporate artistic inquiry methods into its approaches. Today the Exploratorium Tinkering Studio and Center for Art & Inquiry[17] continue to work with artists to explore how artistic methods can be adapted to hands on learning activities for both formal and informal learning contexts.)

In 1974 Douglas Hollis became an artist-in-residence. Working collaboratively with Oppenheimer in the Exploratorium's machine shop,[18] Hollis developed *Aeolian Harp*, a wind-activated acoustic sculpture sited on the Exploratorium roof and over its main entrance. Hollis, who would go on to become a public artist, regularly returned to seek advice from Exploratorium engineers, and today credits the Exploratorium for influencing the investigatory methodology that he has adopted as an artist. The influence was mutual (Exploratorium, 2017).

Hollis' work marked the beginnings of the Exploratorium's development of site-based work in outdoor civic spaces. In the 1980s, as public art and artworks on an environmental scale proliferated, the Exploratorium developed projects such as Ruth Wollan's *The Sea as Sculptress* (1980), Bill Maxwell's *Pandora* (1988) and Peter Richard and George Gonzales' *Wave Organ* (Exploratorium, 1986). Today the Exploratorium continues to develop site-based artworks such as Fujiko Nakaya's *Fog Bridge* (2013) through the *Over the Water* program (Exploratorium, 2018e) and other contexts.

Oppenheimer sought advice from the design duo, filmmakers Charles and Ray Eames, as well as filmmaker Stan Brakhage on how best to present film at the Exploratorium.[19] In the early 1980s he engaged filmmaker Liz Keim to develop an adventurous weekly cinema program. Keim's unique approach evolved into a platform – rare in public education contexts – that furthers exploration of interdisciplinary themes through cinematic experiences developed by independent artists (Exploratorium, 2018f).

Oppenheimer was interested in investigating artistic processes (Hein, 1990, p. 167). In the 1980s the Exploratorium developed the program *Speaking of Music*, a public forum hosted by experimental music curator Charles Amirkhanian. The program format combined performance with conversation; and featured musicians, sound artists, and composers. Today the program has been revived as the *Resonance* series (Exploratorium, 2018g). Another current public conversation series *Ways of Knowing*, explores the creative practices of artists, scientists, and others. Over time, we hope to learn more about the creative process by talking to a diverse range of thinkers and doers, and documenting their approaches and methods. By shedding light on artistic process, we hope to validate the instincts and intuitive tendencies of our publics.

Exploratorium arts philosophy

The Exploratorium's inquiry-based, experiential learning philosophy finds its roots in progressive education, particularly the ideas of American philosopher John Dewey. In recent years, we have embraced the notion of art as a form of inquiry.

Dewey advocated for the democratisation of art, believing that everyone benefits from learning with and through the arts (Dewey, 1934). Dewey directly informed the approach of the celebrated Black Mountain College, the influential, experimental liberal arts school in North Carolina. Its founder Andrew Rice understood how an art-based education could transform society by engendering 'not professional painters, sculptors, musicians or writers of novels and poetry, but people who will have the artistic approach to life as a whole and to everything in life' (Adamic, 1936). This holistic, egalitarian vision of an art education for everyone underpins the Exploratorium's approach.

In 2011 we convened the conference *Art as a Way of Knowing* (Exploratorium, 2011). The conference provided the opportunity to take a fresh look at the Exploratorium's role in the wider fields of science, art and education. The conference proved pivotal since it led to a re-articulation of the Exploratorium arts philosophy as "art as a way of knowing". This subtle re-framing has been revelatory to our work and focused our efforts on better understanding how the methods of art, like those of science, are forms of inquiry (McDougall, 2013).

Upon re-locating to our current site on the San Francisco waterfront in 2013, we formed the Center for Art & Inquiry, a laboratory for the arts within the wider learning laboratory of the Exploratorium (Exploratorium, 2018h). Our recent work has increased our conviction that approaching "art as inquiry", has the power to break through epistemological complexities towards building an enduring rationale for the arts in public learning.[20]

While the pursuit of art encompasses "not knowing", going down unknown paths, becoming lost in murky territory, and embracing ambiguity; it can also lead to insight, discovery and new forms of knowledge.[21]

Approaching art as a verb,[22] as a series of gerunds, begins to capture how art can be "a way of knowing". For example, artists working in the context of the Exploratorium Bay Observatory[23] in the creation of public exhibits seamlessly meld methods of the observational sciences with the environmental humanities: observing, collecting, mapping, visualising, describing, modelling and so forth. Here artists' projects have taken the form of charts of the atmosphere (Amy Balkin); aerial shoreline studies (Center for Land Use Interpretation); and atlases of the local landscape (Rick and Megan Prelinger). These artworks illustrate how modes of observation, and methods honed through the arts can dovetail with investigatory processes from other fields – such as urban geography, meteorology, astronomy or ecology – to bring us into a better understanding of the complex dynamics of place.

As modes of investigation begin to merge, at a certain point the categories of art and science seem to drop away as we consider ways of imagining, describing and comprehending in general. As artist Robert Irwin stated: 'The object of art may be to seek the elimination of the necessity of it' (Weschler, 1982, p. 128). Ultimately, art serves the critical role of opening us to a wider awareness.

Lawrence Weschler, the keynote speaker at our *Art as a Way of Knowing* conference has made important contributions to illuminating the connection between art and inquiry in his observant descriptions of artistic process. Weschler's longitudinal studies of artists Robert Irwin, *Seeing is Forgetting the Name of the Thing One Sees* (Weschler, 1982); and David Hockney, *True to Life* (Weschler, 2008), illustrate how art can address profound questions worthy of serious pursuit over a lifetime.

In *Seeing is Forgetting*, Weschler reveals what the contours of a life dedicated to artistic inquiry can encompass. The chapters of the book unfold like the course of Irwin's life: one phase of exploration succeeding another. Irwin's work takes on different forms as the nature of his investigation into "phenomenal presence" shifts over time (Weschler, 1982, p. 61). At a certain point in his trajectory, Irwin explains, 'I was becoming much more of a question person than an answer person' (Weschler, 1982, p. 86). New experiences spark new insights, and the queries generated by one series of work leads to the responses in the next. Weschler depicts a sequence of vivid scenes of Irwin: fastidiously creating his famous dot paintings in his Venice studio; thoughtfully absorbing the vast literature of phenomenology (Husserl, Hegel, James, Merleau-Ponty, etc.) on a southern California college campus; and solitarily contemplating the light and creating non-material artworks in the Mojave Desert. *Seeing is Forgetting* suggests how our lives might unfold, richly, intuitively and idiosyncratically when we attend to what surrounds us and pursue a line of thought, whether we become professional artists or go down other paths.

We support the open-ended investigations of artists with the idea that their queries will become distilled in artworks and experiences that in turn engender the questions and insights of our visitors.

We also design experiences that allow direct ways of learning through the arts for our publics.

Interdisciplinary culture and arts stewardship

The Exploratorium has cultivated an environment conducive to interdisciplinary cross-pollination. The success of its arts program relies on an institutional culture adept at bridging discourses.

In his introduction to C. P. Snow's *The Two Cultures*, Stefan Collini urges: 'We need to encourage the growth of the intellectual equivalent of bilingualism, a capacity not only to exercise the language of our respective specialisms, but also to attend to, learn from and eventually contribute to wider cultural conversations' (Snow, 1998, p. lvii).

The Exploratorium internal staff and our external collaborators represent a multitude of disciplines.[24] Drawn together towards the common cause of activating public learning, individuals with wide ranging backgrounds have developed a shared language and set of values that continues to evolve.

Curatorial staff, who understand the Exploratorium's context and have internalised its cross-disciplinary approach and institutional ethos, play a critical role in creating fruitful conditions for successful artistic engagements. Often serving as translators between education, science and art discourses, arts program staff work to ensure that the institution stay abreast of leading edge practices and help guide the project development process.

We seek artists who are curious and inquisitive; who embrace the collaborative nature and populist spirit of the Exploratorium; and who are comfortable bridging multiple disciplines. In addition to looking for artists who are able to develop playful works with wide public appeal, we also value the participation of artists successful at addressing some of the most pressing issues of our time – from climate change to social identity – and their skills in designing public experiences that make complex notions (such as the carbon cycle) comprehensible to visitors.

Arts staff steward projects by making connections to expertise in both the internal and external Exploratorium communities – identifying the talents of sound engineers, atmospheric research scientists, maritime historians and metal fabricators, for example. Each artist-in-residence project benefits from a custom-made team, and the appropriate infrastructure to support it.

Artists receive technical support towards the realisation of ambitious new projects, and often are able to push their work into new directions by working collaboratively with staff engineers, exhibit developers, fabricators, writers, designers or videographers. Artworks presented on the public floor need to be robust to withstand the rigors of the Exploratorium interactive space. We carefully assess whether an artist has the stamina (and the proper support) to undergo a full prototyping and iterative design process.

Context affects the ways in which artworks are approached and understood. Resembling a public classroom more than a white box gallery environment, the Exploratorium is not the most conducive setting for presenting the nuance of artistic voice. Oppenheimer described the importance of an Exploratorium artwork serving as a portal for shared discovery rather than calling attention to the genius of its own making. Artists who gravitate to the Exploratorium, value the opportunity of serving the Exploratorium's educational mission, and understand that the subtleties of their aesthetic approach may become subsumed by the greater institutional project.

While works developed by artists are labelled with their names, artworks complement and often blend in with other exhibits. Most artists that we engage (and many art professionals) value the Exploratorium's accessibility, and the wide public that it reaches.

Implications for public learning

Today international contemporary art has evolved into an expansive field encompassing vast possibilities for artistic practice – from material explorations, to conceptual investigations, to projects designed to bring about social change. Yet, most educational approaches generally deny learners the pleasure of being surprised, compelled, delighted, and moved through the modalities of the arts.

A key challenge is that the term "art" remains strongly associated with fine art mastery. In order to effectively advocate for interdisciplinary arts integration in science centres, and public learning environments, we require a clearer shared picture of the diverse forms that art takes today, and how art can inspire creative learning across a wide range of contexts for people of all ages.

"Art" can be a confusing term since it conveys both the process of making art and its products. The word "art", particularly in its noun form, continues to evoke fine art objects (such as painting, sculpture, or drawing); and the refinement of craft and technique. While the learning dispositions, and aesthetic sensibilities developed through the arts, whether acting, dancing, writing poetry or playing music are invaluable – and can certainly be employed towards interdisciplinary learning – traditional skills gleaned through fine art training are not necessarily required for integrated art learning. The most simple art approaches can easily complement the methodologies of science and humanities to heighten individual meaning making, and open whole new interest areas to learners. Artistic inquiries can theoretically be designed for, and pursued by, any learner of any age in any setting without prior experience in the arts. While working with artists is ideal, it's also possible for curators and educators to interweave art experiences into interdisciplinary learning without directly engaging artists.

The world arrives to us whole – not parsed into disciplinary categories. Ideally we're able to tap into a wide range of perspectives to further our understanding. Holistic learning approaches that meld, interweave and dovetail the methods of the arts with the approaches of other disciplines, have long proven to be powerful ways of drawing upon the full breadth of our cognitive and emotional capacities as learners.

We need to find ways to encourage the inner artist in each of us towards passionate, lifelong learning. We believe that approaching "art as a way of knowing" holds enormous transformational potential for science centres, and public education in general.

In our current era when an increasingly complex world can instill a sense of fear and helplessness; and lead to feelings of individual disempowerment – that intensify a surrender to authoritarian power – an inclusive, international effort to bring diverse communities together towards common understanding, and shared solutions becomes more urgent than ever. The arts are as crucial as they were in Oppenheimer's day toward the vital collective enterprise of public learning and self-determination.

Notes

1 Writer and publisher Dave Eggers is the co-founder of *826 Valencia*, a non-profit writing and tutoring center. He made this statement at the Exploratorium 2015 Gala.
2 Oppenheimer's wife Jacquenette was the Exploratorium's co-founder.
3 Oppenheimer's expertise was in uranium isotope separation.
4 The work of PSSC and ESS reached millions of students and was produced under the umbrella of the Education Development Center. For more see: www.edc.org/history (Accessed 19 May 2018).
5 PSSC and ESS cross-disciplinary teams generated educational materials including workbooks, learning kits, and scores of educational films. Other notable contributors included artist and educator Phylis Morrison; physicist Phillip Morrison; and educator Eleanor Duckworth.
6 During this same period Oppenheimer developed, the "library of experiments" at the University of Colorado. Nearly 100 physical models demonstrating classic physics experiments allowed students to freely explore phenomena at their own pace driven by individual curiosity. Science educators praised Oppenheimer's library since it brought to life abstract concepts, otherwise described in textbooks. The library became replicated in other university settings, and today is viewed as an important precursor to the Exploratorium.
7 On this trip Oppenheimer visited museums including the Palais de la Découverte in Paris, the Deutsches Museum in Munich, and the Museum of Science and Technology in Tel Aviv.
8 Designed by Bernard Maybeck, *The Palace of Fine Arts* was originally built for the Panama-Pacific International Exposition of 1915.
9 NPR Science Friday Profiling Frank Oppenheimer *Originally broadcast Aug. 7, 2009* (Radio interview with K.C. Cole).
10 The exhibition had travelled to the Corcoran Annex at the Smithsonian Institution in Washington D.C. before its presentation at the Exploratorium. See Hein, 1990, pp. 33–39.
11 The Exploratorium collection includes *Drawing Board* (by John Ravilious); *Albert* (by John Billingsley); a series of *Harmonograms* (by Ivan Moscovich); *Dioxiimorekinesis* (by Irving John Good and Martine Vite); and a highly modified version of *After Image* (by Christopher Evans).
12 In 2011 the Exploratorium commissioned artist Fujiko Nakaya to create *Fog Bridge*, an ephemeral sculpture. The technology behind the work was developed

for the 1970's Osaka Pepsi Pavilion by Nakaya's E.A.T. collaborator, cloud physicist Thomas Mee.

13 Kepes collaborated with physicist William Walton, artist Mauricio Bueno, and electronic music composer Paul Earls in the creation of the work.

14 Hein notes: the 'grant marked the NEA's acknowledgement of the aesthetic dimensions of science, and it enhanced the legitimacy of science centers as museums, a status that had not been accorded them by the American Association of Museums' (Hein, 1990, p. 152).

15 Artist Peter Richards, a friend of the Oppenheimer family in Colorado, would eventually become an Exploratorium Senior Artist and director of the arts programs.

16 School in the Exploratorium is now called the Institute for Inquiry.

17 The recent Center for Art & Inquiry project *Begin Here: Assignments, Instructions, Prompts and Cues from the Artist's Classroom,* developed with collaborators from the Josef and Anni Albers Foundation and the Rhode Island School of Design, explores exercises developed by artists conducive to interdisciplinary learning. A long history of approaches developed by teaching artists requires further research, and mining for contemporary contexts. (Exploratorium, 2018d).

18 Open to public view – the machine shop historically served as the heart of the Exploratorium – and became a hub of activity for an extended community of artists like Hollis. As the Exploratorium has evolved, other spaces such as the biology lab, the Tinkering Studio, the Bay Observatory, and the Studio for Public Space have become creative engines for the Exploratorium's R&D work.

19 In 1974 Jon Boorstin produced the short film *Exploratorium* with funding from both the National Science Foundation and the National Endowment for the Arts.

20 See Exploratorium (2018i) and McDougall (2015) for descriptions and documentation of recent projects.

21 The way "discoveries" or "knowledge" is described and defined in the arts and humanities is quite different than in the sciences and requires much more discussion than possible here. For example, in art/science contexts in the U.S., Robert Root-Bernstein presents case studies that document how artistic explorations are connected to scientific breakthroughs and invention (Root-Bernstein, 2018). While these case studies are compelling, there is a well-known danger of instrumentalising the arts. Art speaks its own languages, and the advances of the arts to cultural "knowledge" should be seen as valid in their own right, and of serious consequence.

22 Richard Serra's artwork *Verblist* (1967–1968), a written list of infinitives including: "to roll", "to crease", "to gather", "to delete", "to modulate", etc., provides a window onto the wide – ranging activities of one artist's process.

23 Exploratorium senior artist and curator Susan Schwartzenberg directs the work of the Fisher Bay Observatory and co-directs its environmental initiative (Exploratorium, 2018j).

24 Some embrace the term *non-disciplinary*.

Bibliography

Adamic, L. (1936) Education on a Mountain: The Story of Black Mountain College, *Harper's Magazine*, April, 519.

Cole, K.C. (2009) *Something Incredibly Wonderful Happens: Frank Oppenheimer and His Astonishing Exploratorium*, Boston and New York: Houghton Mifflin Harcourt.

Cybernetic Serendipity. (n.d.) *Cybernetic Serendipity* [Online]. Available at: http://cyberneticserendipity.net (Accessed 19 May 2018).

Dewey, J. (1934) *Art as Experience*, New York: Berkeley Publishing.

Exploratorium. (2018a) *Bob Miller/Light Walk* [Online]. Available at: www.exploratorium.edu/arts/begin-here/bob-miller-light-walk (Accessed 19 May 2018).

Exploratorium. (2018b) *Ruth Asawa* [Online]. Available at: www.exploratorium.edu/arts/artists/ruth_asawa (Accessed 19 May 2018).

Exploratorium. (2018c) *Ruth Asawa/Milk Carton Sculpture* [Online]. Available at: www.exploratorium.edu/arts/begin-here/ruth-asawa (Accessed 19 May 2018).

Exploratorium. (2018d) *Begin Here* [Online]. Available at: www.exploratorium.edu/arts/begin-here (Accessed 19 May 2018).

Exploratorium. (2018e) *Over the Water* [Online]. Available at: www.exploratorium.edu/arts/over-the-water (Accessed 19 May 2018).

Exploratorium. (2018f) *The Cinema Arts Program* [Online]. Available at: www.exploratorium.edu/arts/cinema-arts (Accessed 19 May 2018).

Exploratorium. (2018g) *Resonance* [Online]. Available at: www.exploratorium.edu/arts/resonance (Accessed 19 May 2018).

Exploratorium. (2018h) *Center for Art & Inquiry* [Online]. Available at: www.exploratorium.edu/arts/center-art-enquiry (Accessed 19 May 2018).

Exploratorium. (2018i) *Arts at the Exploratorium* [Online]. Available at: www.exploratorium.edu/arts (Accessed 19 May 2018).

Exploratorium. (2018j) *Fisher Bay Observatory Gallery 6: Observing Landscapes* [Online]. Available at: www.exploratorium.edu/visit/bay-observatory-gallery (Accessed 19 May 2018).

Exploratorium. (2017) *Archimedes* [Online]. Available at: www.exploratorium.edu/exhibit-archimedes (Accessed 19 May 2018).

Exploratorium. (2011) *Art as a Way of Knowing* [Online]. Available at: www.exploratorium.edu/knowing/ (Accessed 19 May 2018).

Exploratorium. (1986) *The Wave Organ* [Online]. Available at: www.exploratorium.edu/visit/wave-organ (Accessed 19 May 2018).

Hein, H. (1990) *The Exploratorium: The Museum as Laboratory*, Washington and London: Smithsonian Institution Press.

McDougall, M. (2015) Artistic Inquiries, Public Inquiries, *Art Practical* [Online], 7(2), 29 October. Available at: www.artpractical.com/feature/artistic-inquiries-public-inquiries/ (Accessed 19 May 2018).

McDougall, M. (2013) Exploratorium as Laboratory and Studio, *Arcade* [Online], 31(3), Summer. Available at: http://arcadenw.org/article/exploratorium-as-laboratory-and-studio (Accessed 19 May 2018).

MIT's Center for Advanced Visual Studies. (n.d.) *Center for Advanced Visual Studies Special Collection* [Online]. Available at: http://act.mit.edu/cavs/ (Accessed 19 May 2018).

Oppenheimer, F. (1977) *The Arts: A Decent Respect for Taste, The National Elementary Principal* [Online], 57, 1, October. Available at: www.exploratorium.edu/files/about/our_story/history/frank/pdfs/respect_for_taste.pdf (Accessed 19 May 2018).

Oppenheimer, F. (1976) *Growing Up in the Arts, The National Elementary Principal* [Online], 56, 1, September–October. Available at: www.exploratorium.edu/files/about/our_story/history/frank/pdfs/growing_up.pdf (Accessed 19 May 2018).

Root-Bernstein, R. (2018) How Arts, Crafts and Design Training Benefit STEMM Professionals: The Evidence and Its Limitations, in: *Proceedings of the NAS Sackler Colloquium Creativity & Collaboration: Revisiting Cybernetic Serendipity*, 13–14 March.

Snow, C.P. (1998) *The Two Cultures*, Cambridge: Cambridge University Press.

Weschler, L. (2008) *True to Life: Twenty-Five Years of Conversations with David Hockney*, Berkeley and Los Angeles: University of California Press.

Weschler, L. (1982) *Seeing Is Forgetting the Name of the Thing One Sees: A Life of Contemporary Artist Robert Irwin*, Berkeley and Los Angeles: University of California Press.

5.2 Art thinking

Christopher Lindinger and Maria Pfeifer

English Translation: Mel Greenwald

How do we actually want to live?

According to what you read in the papers, we live in a time of huge technological advances, and it is said that these developments will have unprecedented impacts on human society – on our coexistence with others, life in the workplace and the world of education. Hardly a day goes by without a print publication or website enumerating the various occupations that will be superfluous in the future – though there has not been total agreement among these lists heretofore.[1] The feeling that we live in a world in which there is an apparently unstoppable dynamism in the emergence of new findings dovetails nicely with the fact that technologies are developing faster than we are understanding how to use them. In order to compensate for this incongruence, we have to get away from a technological approach to this issue and frame it in social terms. We as a society have to make every effort to answer a simple question: How do we actually want to live?

The fact is that digitisation and automation are no longer futuristic visions. Industry 4.0, AI, the Internet of Things, robots and virtual reality are the here-and-now; ignoring them will not make them go away. Once technological developments have been unleashed on the free market, they cannot be recalled. Regardless of whether you look at Moore's Law, trend reports or predictions from the past, the curves are moving steadily higher. What they tell us is that these developments will continue at increasing speed. This technological progress is mirrored by a social discourse that is fraught with fears. But the core problem is not how these developments could be stopped but rather how we can proactively configure them. After all, whether we benefit from these developments or they prove to be our undoing is not a matter of technical feasibility but rather a sociopolitical question. This is why we have to engage in a sociopolitical discourse that gets people in all walks of life and all interest groups involved in facing this issue – the question of how we want to live; the question of whether everything that can be automated should be automated; the question of what constitutes a desirable future, one worth living.

Seeing to it that this discourse can happen calls for taking numerous steps and assembling a bundle of strategies. Amongst this multiplicity of strategies in this discursive process, a very important role is attributable to art. In all technical and social developments, art has always played a relevant part – as visionary, as source of inspiration or as social corrective. The exhibition space, a transitional zone between public and private domains, is a setting in which, with the right strategy, there prevail ideal conditions to foster this discourse with the help of art. Museums in general and science centers in particular identify themselves as places where this corrective can take place – or, for that matter, must take place. In addition to a museum's classical responsibilities – conserving, collecting, conducting research – nurturing a discourse, conveying socially relevant content, is one of this institution's most important tasks. 'Museums have an important duty to develop their educational role and attract wider audiences from the community, locality, or group they serve' (International Council of Museums, 2017). The mission of science centres and educational museums in general includes bringing as many segments of society as possible into contact with science and current research. They build bridges to fields – research labs and elite universities – that are otherwise inaccessible to large parts of society. Science museums want to enable us, their visitors, to experience the seemingly incredible findings of science, fascinate us with interactive exhibits, and awaken our enthusiasm for and understanding of research. Museums want to mediate encounters with science; that is why they are ideal locations at which to stage such social discourses. It is an absolute must that all museums face the challenge of not only imparting knowledge but also listening to visitors and, based on this feedback, renegotiate entrenched models of knowledge production and research. Open access, gender equality, ethics and democracy, commitment and participation cannot just be distant ideals; they have to be actively embedded. In this context, museums can function as role models – as extramural educational facilities that provide open, permeable interfaces between broad segments of society and first-class scientific expertise.

Museums worldwide have come up with ways and means to face these challenges, and their approaches are often outstanding and innovative – open labs, workshop programs, hands-on exhibits, experimental structures and interactive installations; with historical artefacts from scientific workshops, speeches, hack-a-thons and science cafés . . . Indeed, there are lots of ways to propagate science and to enable the general public to participate. In 2017, Ecsites' Spokes Magazine summarised the relevant future trends and tasks for science centers, whereby the highest priority is to also make a (socio)political statement, to take a position against relativism, and to enable visitors to develop competence dealing with media and facts. The aim is to not just remain entrenched on the "commitment" level but rather to initiate a more in-depth discussion (Okkersen & Becker, 2017).

Impact

> The inability to imagine a world in which things are different is evidence only of poor imagination, not of the impossibility of change.
>
> (Bregman & Manton, 2017)

Framing refers to the interpretational frameworks that we apply to our thoughts and actions, our behavior. These interpretational frameworks influence how we select, classify, contextualise, weight and interpret information. The collective framework/interpretational horizon is always established by the strongest social protagonists – in the case of digitisation hype, for instance, by tech corporations, industries and other profit-oriented institutions. In combination with the existing media discourse, they establish the collective framing. When you follow the discourse on AI and robots in the media, it sometimes seems as if the abolition of humankind is looming on the horizon. Moving about within this interpretational framework, it quickly appears that there is only a single direction in which the world can develop, one route that progress can take. Within this interpretational framework, we are reduced to mere passengers permitted to move along predetermined paths. Individuals come to see themselves as incapable of taking independent action and, accordingly, withdraw into the private sphere and abjure responsibility. To escape from this deterministic world view, we have to break through the frame, call it into question and re-establish it.

So much for theory. In real life, however, it is not easy for people to radically change their point of view and to consider a product or a technology from another perspective. Nevertheless, this is precisely where we find the extraordinary power of art, which has the capability to bring about these shifts in perspective – on intellectual, emotional or aesthetic levels. Since art is always ambiguous, art projects can depict the complexity of science, the contradictions that are inherent in a technology or scientific breakthrough, and the uncertainty that goes hand in hand with visions of the future. Spotlighting developments, making a fuss when things are going awry, questioning the status quo, encouraging people to ponder the way things are, to focus on things we take for granted, and take them to the absurd – art works with all of these strategies.

If we museums, science centres and exhibition venues want to be places of critical education, and our goal is to host a critical, socially relevant discourse with our visitors, we can make art our ally. The involvement of artistic strategies and/or artists as individuals in the conception and design of exhibitions and the public's encounters with their content rewards us with exhibitions, objects and activities that appeal to visitors aesthetically (though not exclusively so). That does not necessarily mean that they have to be beautiful, but rather that they communicate via their form/design.[2] In contradistinction to theoretical texts and factual explanations, artistic artefacts speak to visitors on sensory and emotional levels. In a society that is increasingly described as post-factual, this is a welcome alternative way to get people to proceed to consider a topic and subsequently to also sustainably engage with it. A particular product may be quite "sensible and logical" within the general interpretational framework, but amidst another one it quickly appears to be absurd. Conversely, a work of art might come across as impossible or utopian at first glance, but a slight reassessment of the status quo and the application of some common sense could suddenly make it appear to be only one of many possibilities. And all of a sudden, this diversity liberates us from following pathways that had previously been so clearly delineated and opens up latitude for action.

These changes – re-framings – do not occur overnight. Nevertheless, artistic projects present us with the possibility to talk alternatively about the developments of our time, to describe its problems differently and to rethink solutions to these problems. Thus begins progress – and, indeed, progress as the opposite of persistence due to paralysis. Because there is one thing we cannot forget: technology, digitisation, automation and artificial intelligence are all things made by humankind.

The artistic view of the future has especially high potential to formulate solutions to problems that have not even emerged yet. Furthermore, the art context offers the freedom to deliberately, playfully confuse spectators, something inherently lacking in a scientific context. Works of art do not have to go through a peer review process or produce hard data. As suggested above, their strength lies in the ability to communicate. By conducting what-if experiments and designing a solution for a problem based on the assumption that a futuristic technology already exists, artistic strategies can call into question assumptions and preconceptions about the role of these technologies in everyday life. By bringing forth objective manifestations and visualisations of future scenarios, artists make available the experience of these possible futures. These experiences enable us to confront problems before a product comes onto the market and becomes omnipresent.

Creating an actual object, a concrete rendering, a prototype leads to questions that a purely theoretical encounter simply cannot bring out, questions that might previously have been undreamt of. The sensory experience in an exhibition context turns visitors, on short notice, into experts who can have recourse to the immediate experience and formulate conclusions on that basis. These assertions and impressions can then be brought up and contextualised by personnel trained to mediate visitors' encounters with exhibition content, whereby, in doing so, it is essential to bring in the specific reality of visitors' lives. This is why it is indispensable to continually consider how we can lower the threshold and encourage the general public to partake of our offerings – to reduce inhibitions on the part of various groups to spend more time in our facilities, as well as to reduce the individuals' inhibitions to then go out into the world. The essential question is how we as cultural and scientific intermediaries, as exhibition designers and curators, as scholars and teachers can help people to go into the world and do something, how to enable and strengthen them to get active on their own, to take their lives into their own hands and to actively influence technological, social and scientific development.

Art thinking strategies

> I do not believe in Art – I believe in Artists.
>
> Marcel Duchamp (in Tomkins, 1996, p. 408)

There has been a lot of talk about art and science for several years now. What has, in fact, been subsumed under these buzzwords – as this chapter will now demonstrate – is a multiplicity of strategies and various modes of dovetailing art and science. Our many years of experience have yielded the following four strategies that have proven to be especially effective and reasonable ways of initiating a critical discourse on scientific topics with the general public: artistic intervention, artist collaboration, future ideation and speculative intent (see Figure 5.2).

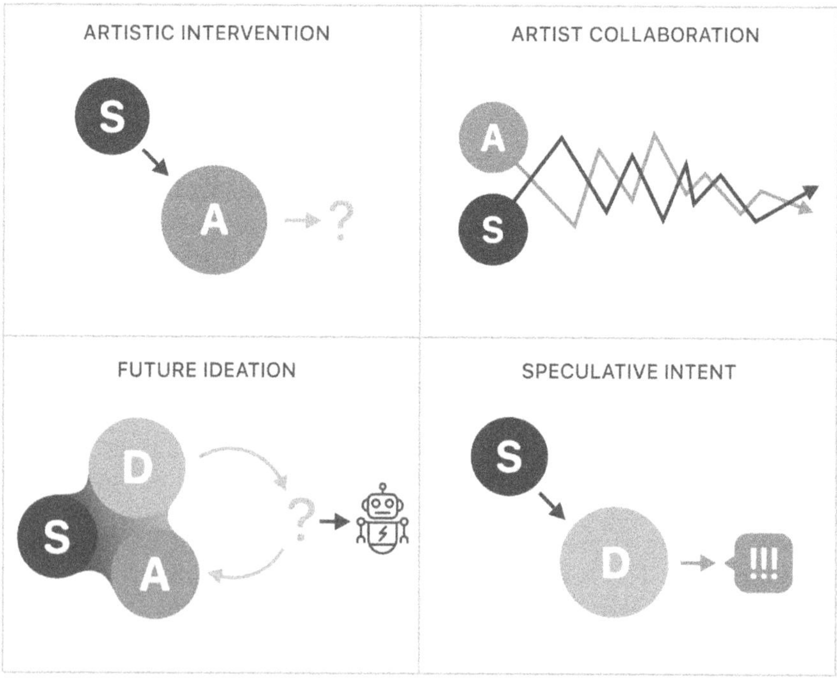

Figure 5.2 Art Thinking Strategies

First art thinking strategy: artistic intervention

In an artistic intervention having to do with science, artists take up and deal with scientific topics in their work, whereby direct exchange between the artist and scientists does not necessarily have to occur. The result is a work of art that treats provocative issues. This provocation and the multiplicity of meanings inherent in the work of art lead to discussion.

Artistic works communicate primarily on the emotional-sensory – and not on the cognitive – level. Artistic objects thus enable those beholding them to grasp science, theory and abstract technological developments; they make them comprehensible with the senses. Thus, we can reach broad segments of the population on a level on which specialised knowledge is not a prerequisite. Objects on exhibit can often be excellent ways to approach and treat a broad spectrum of topics, since objects and actions successfully amalgamate numerous aspects in a single image. The aesthetically fascinating object that shows scientific/technological aspects from a novel perspective constitutes an invitation to engage in a factual, critical discourse. In addition to this aesthetic appeal to exhibition visitors, works of art have the inherent quality of being ambiguous, equivocal. This might cause alarms to go off in the minds of some scientists – Where's the hard data? – but there is really no reason to get up in arms! In fact, it is precisely this multiplicity of meanings that is the outstanding quality of works of art, particularly in the context of an educational museum – the hook that holds a beholder's attention and leads to the formulation of a question – one that is conducive to initiating communication between visitors and the guides on duty in our exhibitions. This suddenly induced need to pause and reflect is important. After all, art's power is, to no small extent, attributable to its capacity to wrench things out of their customary contexts and serve them up again with a twist. The upshot is a sort of annoyance conducive to rethinking.

Second art thinking strategy: artist collaboration

In this constellation, artists and scientists encounter one another at eye level as experts who are interested in what their counterparts are up to, and who want to get involved in collaborative processes. Two different systems of reception and discourse engage in exchange and mutual reference, and begin renegotiating what were thought to have been matters taken completely for granted. The core of this strategy is the process of exchange, not the final result, the reciprocal movement between the disciplines that, ideally, makes possible a shift in perspective on the part of science as well as of art.

The artistic output in this case is an accessory that can subsequently function as a tool for getting across scientific topics. Likewise, a work's scientific background can serve to impart to another group an understanding for graphic arts. The intermediary approach taken by art and science projects is a matter of vice versa: art gets across science; scientific aspects convey art.

A widespread practice used to foster such collaborations is to set up framework conditions that enable artists to spend residencies at leading-edge scientific institutions – for instance, the Collide International Residency at CERN in Switzerland, and the artistic residencies at the European Space Agency and many other smaller facilities that are often staged in cooperation with partners in the world of art and in cultural fields. The results – the work produced thereby – can then be presented in both a scientific context (science communication, conference, institute, etc.) as well as in a museum. Behind this is also a general desire to bring science and society closer together. Efforts to integrate society into matters of science and innovation, to negotiate them in public and to democratise decision-making can be supported by art-science collaboration. Artists-in-residence at scientific facilities gain an outsider's point of view of the research taking place there, and can help to identify new areas of impact or consequences for society. The necessity for scientists to communicate their research results to "outsiders" supports the process of perspective shift. In the role of a transdisciplinary "wild card," the artist can demonstrate interrelations between research topics that normally would not be associated with each other. Furthermore, this strategy can provide traction to scientific processes and research systems that have gotten bogged down, call their assumptions into question, and make them more accessible by broad segments of society.

Third art thinking strategy: future ideation

Future ideation has to do with inventiveness and designing prototypes of all sorts. Its objective is to allow spectators to behold how things should be and not to depict how they are. This makes it an ideal framework for the investigation of future scientific scenarios and new technologies. Interdisciplinary teams made up of designers, artists, scientists and engineers jointly elaborate possible technical and social innovations. These are meant to yield desirable and, at some point, feasible visions of the future.

A combination of artistic, scientific and technical expertise makes it possible to do innovative brainstorming, create sketches, produce artistic prototypes or design studies, and specify how developments in science and technology might influence society. By means of prototypes, our conceptions of the future can be subjected to well-founded testing and thus develop sustainable strategies to have input into how that future turns out. In contrast to trend research, this is not about predicting, on the basis of economic-capitalist considerations, which horse to bet on in this race; rather, the point is to reflect on current technological developments – so-called progress in science and technology – and to actively influence it by engendering innovations.[3]

Preoccupation with the future – not with far-distant futures but rather with a future that seems to be on the cusp of becoming the present – is a practice that, at first glance, might seem somewhat out of place in the field of classic museology. On the other hand, it is standard operating procedure for museums to present and convey historical developments in a way that relates them to the present. In addition to conserving cultural and natural historical artefacts, many museums and similar institutions see it as a part of their mission to get a glimpse of what lies ahead. Precisely because they often have a well-researched historical collection at their disposal, they – standing on the metaphorical shoulders of giants[4] – are in an ideal position to see a preview of what we are in for. In educational museums, appropriate approaches include integrating this strategy into workshops and other participative formats, or into projects involving an in-house R&D division or independent associates. The results can be presented at events involving theoretical elaborations or exhibitions in gallery spaces. This strategy is also superbly suited to bringing in private-sector protagonists who are often difficult to reach with other commitment strategies – and, after all, corporations as the producers and creators of the actual things with which we configure our everyday life play a relevant role when it comes to questions of configuring the future.

Fourth art thinking strategy: speculative intent

In contrast to future ideation in which the focus is first and foremost on responsible, desirable innovations, speculative intentions are approaches that work with jarring, counterintuitive creations cleverly designed to shake people up a bit. This strategy employs products that seem genuine but will never be marketed. The process resembles that of science-related artistic interventions in which designers and/or artists seize upon a scientific or technological innovation, extrapolate it to the next conceptual level, and then associate it with a quotidian object or envision this futuristic technology in the form of a fictitious product inserted into a common, everyday setting. Here, all the ways and means of product and graphic design are deployed to conjure up absurd futuristic products that toy with visitors' horizon of expectations and, briefly, fool them. As with artistic interventions, the intended result is, once the irritation and indignation subside, to open up a process of questioning. In any case, this is more of an alternative future scenario we can compare with the reality of our own lives.

As previously noted, it is no easy task to change future expectations, to modify frames and interpretational patterns, or at least to make them recognizable as such. Speculative strategies conceive and execute supposedly real products, most of which have something provocative or even disturbing about them, and inserts them into familiar contexts. This discipline is not out to invent successful, innovative products; instead, the mission is to question precisely this product landscape and to challenge the discourse prevailing in it. In his article *Speculative design: crafting the speculation*, J. Auger describes in the following terms the effect that best-practice speculative methods can have on beholders:

> The examples described are therefore intended to present [. . .] how it must be managed and crafted to connect to a specific audience's perception of the temporal world around them. Once established these perceptions can be stretched or manipulated in precise and informed ways. These exist as plausible, tangible and accessible demonstrations, or more specifically hypothetical translations of disruptive technological innovations into the future products they could become. The key benefit of this approach is the removal of the commercial constraints that normally direct the creative process. This decoupling allows for the goals to be based on questions and discourse rather than market-led agendas; hypothetical possibilities not real products; utopian concepts and dystopian counter-products. They can inspire an audience to think not only about what they do want for their future selves but also what they do not want.
>
> (Auger, 2013)

These objects are thus highly provocative, conceptual and dedicated to showcasing problems in terms of their solutions (cf. Dunne & Raby, 2014). Taking this approach in an exhibition context at a science museum yields a variety of possibilities. The purported product conveys visitors away from a concept that is speculative and rather difficult to grasp, and confronts them with a concrete, haptic, seemingly everyday thing that, at first glance, seems to be something currently available from nearby retailers, something they either like or do not. In this way, they make a transition from observers to people actually affected by this phenomenon. This strategy does not have to be employed in an exhibition context; it is also appropriate for use online or in other media-based actions. And since it plays with the principles of advertising and the market economy, it is also superbly suited to gain the attention of and provoke discussions by those who do not frequent museums.

Conclusions: everything will be OK!

If we strive to live up to a mandate for museums to function as discursive venues, then art is an effective strategy to do so. Artist collaborations, speculative intent, future ideation and artistic interventions are means of integrating a broad spectrum of social protagonists into the process of science and research and to keep them involved. All of these artistic strategies have the potential to encourage active commitment and to kick off discussions. When we collaborate with artists to imagine a possible future, we are perhaps not in a position to predict the future but we can raise important questions and discuss problems associated with emerging technologies before their social and ethical implications manifest themselves.

The aim is to continue to perceive the world as a place that human beings can configure and to expand humankind's latitude for action. If museums wish to position themselves as indispensable actors in this process, they must display the courage to be the beacons of a discourse that poses the critical questions.

No matter how critical, dangerous, precarious and troubling the future might sometimes seem to us, we nevertheless want to take advantage of this opportunity to demonstrate our support for the utopians. And, in the final analysis, the world has developed in a generally positive direction over the last few decades. Projects like *Our World in Data* by economist Max Roser demonstrate how and under what circumstances our living conditions are changing for the better. In stark contrast to media reportage that focuses on particular events, this online publication by Oxford University describes long-term processes that are changing our world.

> The number of people living in extreme poverty fell from close to 2 billion in 1990 to 0.7 billion in 2015 [. . .]. On no day in this 25 year period was the headline of any newspaper in the world 'The number of people in extreme poverty fell by 137,000 since yesterday'. This is despite the fact that – on average – this would have been an accurate headline every single day during these 25 years.
>
> (Our World in Data, 2018)

Here as well, the objective is a shift in perspective. Not everything is incessantly getting worse. No! But it is up to us to work to make the world a place in which living conditions are improving for all of us.

Notes

1 The MIT Technology Review recently attempted to summarise all studies on this subject. Its conclusion: 'There are about as many opinions as there are experts' (Winick, 2018).
2 Nota bene: Every object in an exhibition does this, regardless of whether it was designed by an artist or not. Each object, every infographic, all furnishings placed in the context of an exhibition are first perceived in sensory terms and only thereafter with respect to their content.
3 Armin Bechmann, in his 2009 book "Der Paradigmensprung: Argumente für eine Neuorientierung des Umganges mit Natur, Leben und Wissen; Programmatik, Konkretisierung, Institutionalisierung", differentiates between, on one hand, future ideation and, on the other, systemic futurology and trend research in the following terms: 'The primary aim of trend research and systemic futurology is to describe possible and expected developments. Thus, they endeavor to help the beneficiaries of their insights adopt modes of behavior adapted to these developments, whereas future ideation strives to point out or create options for future action and thereby to encourage people to actively configure the future' (Bechmann, 2009).
4 'If I have seen further it is by standing on ye shoulders of giants' (Westfall, 1994, p. 106).

Bibliography

Auger, J. (2013) Speculative Design: Crafting the Speculation, *Digital Creativity*, 24(1), 22.

Bechmann, A. (2009) *Der Paradigmensprung*, Barsinghausen: Verl. Ed. Zukunft.

Bregman, R. & Manton, E. (2017) *Utopia for Realists*, 199, New York: Little, Brown and Company.

Dunne, A. & Raby, F. (2014) *Speculative Everything*. [S.l.]: MIT.

International Council of Museums. (2017) *Code of Ethics: ICOM Code of Ethics for Museums*, 24, Paris: International Council of Museums.

Okkersen, M. & Becker, J. (2017) Watching the Trendwatchers, *Spokes* [Online], 27. Available at: www.ecsite.eu/activities-and-services/news-and-publications/digital-spokes/issue-27 (Accessed 31 January 2018).

Our World in Data. (2018) *Our World in Data* [Online]. Available at: https://ourworldindata.org/ (Accessed 01 February 2018).

Tomkins, C. (1996) *Duchamp: A Biography*, New York: H. Holt.

Westfall, R. (1994) *The Life of Isaac Newton (Canto Original Series)*, 106, Cambridge: Cambridge University Press.

Winick, E. (2018) Every Study We Could Find on What Automation Will Do to Jobs, in One Chart, *MIT Technology Review* [Online]. Available at: www.technologyreview.com/s/610005/every-study-we-could-find-on-what-automation-will-do-to-jobs-in-one-chart (Accessed 31 January 2018).

5.3 Art as ingredient for meaningful science learning

Samar D. Kirresh

Introduction

When multiple disciplines are involved, it is hard to precisely communicate one's views. There is a sort of "clear confusion" in the mind, where thoughts overlap and it is difficult to separate what is naturally connected. A set of drawers is often used as an analogy to describe the way our brain organises knowledge from different subject areas. I personally prefer the image of a wardrobe, where items are all hanging together and can be easily shuffled, repositioned and rearranged. Pick a summer shirt and pair it with a raincoat if you want; who can say otherwise?

The mind brings out what is needed in a specific moment, in various combinations, but it still sees it all. At least, this is how I see my brain working when I am involved in art and science educational activities. Both fields of knowledge are present, at the same time, with different items belonging to each one. What comes first, what follows and what is shared depends on the individual learner and is the result of an open and non-restricted design that follows a need, not a rule. We can say that imagination and expression are strong features in art practice, while enquiry and discovery characterise all scientific thinking. Together they form combinations that can support effective learning and expression.

In their research, Tyler and Likova (2012) describe the role of arts in learning and cognition from the perspective of neurosciences. They describe the interrelations between brain sections that seem to be activated when arts and learning are involved. It is thus interesting to think of which educational activities can be designed that might encourage these parallel activations in the brain. How different could these activities be? And what role would science and art each play within the learning experience?

Meaningfulness, participation and democracy

There are many reasons in our local context that confirm that we need combinations between art and science educational activities. School education in Palestine has been based for a long time on teaching subjects as separate, non-integrated disciplines. What prevails is a misleading interpretation of the arts and sciences as separate fields that demand different teaching-learning methods. The system drives students to draw lines between what is considered to be an academic learning path for sciences and a talent-based path for arts.

A person that studies or works in a scientific area is usually believed to be an *academically achieving* student who had the chance to follow an academic learning path. This usually entails more opportunities for university education and better careers. An artist is most of the time thought of as a student who *did not make it* through, yet he/she has *skills and talents* that serve well in battling through life.

At the moment, learners do not have many chances to explore their creativity or use their imagination in proposing personal ideas or discovering new ones. The system ignores possible interactions between science and art that might in fact be functional to the processes of learning itself, encouraging creative and critical thinking or expression. Art lessons are not offered by many schools because of low budgets for the running cost of such classes, scarce provision of art supplies, or in the belief that there is a "better" use of lesson time for other more important subject matters such as languages and science. When implemented, art lessons mostly propose the making of artefacts through step-by-step instructions, with rare opportunities for creativity or self-expression. Science education, on the other hand, is in no better shape. Though science lessons are broadly offered, activities are often limited by traditional test-driven teaching methods and a curriculum that primarily demands memorisation of facts and data. Hands-on opportunities are minimal for the same reasons of low budgets or lack of resources and lab supplies caused by restrictions connected to the territorial Occupation.

As a science-art educator, I ask myself all the time what it means to create art-science learning opportunities and I find a new answer each time. In every activity I design there are different specific nuances, but the overall common denominator seems to be the idea of putting science contents in relation with individuals and with their personalities and lives.

At the beginning I tried to use science as a technique to make art. For example, I used chromatography to create coloured paper. This first naïve approach soon developed into something quite different. I was obsessed by the idea of relating these two fields of knowledge, yet at the same time I was terrified of being contradicted, especially after the many conversations I had with science educators and artists about what the coexistence of art and science might mean for them. Everyone seemed to appreciate my excitement, but they were all very cold about the possibility of experimenting with mixing the two areas of work: science educators did not see how science could open up different scenarios. For them science was to be learned, applied or used for producing more science. Utilising science to get to an artistic product was fine as long as it was not regarded as actually doing or learning science, but rather as an additional fun activity for children. In their view, this was definitely not a learning experience. Artists on the other hand, made me feel like I was violating territory and trespassing on tribal land. Art for them was beyond the abstraction of science. Of course, they recognised that their tools and materials were based on much science and technology, but they considered this as a different "science". It sounded almost as if art could be murdered by science.

With these external judgements in mind, my next steps were careful but daring. Artistic and scientific concepts needed to be harmonised. Each was to serve the other and emphasise it. Activities had to be designed to support expression and learning. And they were made so as to preserve the characteristic physical spaces (places of practice) and virtual spaces (social interactions and relations) of each field. The idea was not to fill the deficiencies of one by using the other, nor to create attention for one at the expense of the other. The challenge was to create an equal space for each one to grow and sustain individual learning processes.

Most experiences engaged participants in creating collaborative art works in science festivals outside the Qattan Foundation Science Studio in Ramallah (where the activity design took place). Activities were organised in informal, open contexts that seemed to encourage people to try things out and engage freely with the proposed activities. These informal contexts appeared to be particularly successful in loosening up tensions and encouraging participants to reveal their own knowledge, skills and ideas. Art and science skills and content were never judged. Skills and materials were simple and encouraged play and exploration, which was successful with children and adults alike, with spaces designed to accommodate a large visitor spectrum. Participants were invited to simply express themselves with basic tools and materials and to enjoy producing artworks with others. The emerging of a common memory of the experience through a social participatory act deeply supported learning.

Materials were simple, also because they had to be available in our closed environment. The design of the experiences offered a new reading of materials and tools from artistic and scientific perspectives. Everyday tools and materials were used in new and different ways. Sand, flour, coffee filters, iron, sanding paper, fabric and food colouring became experimental material and means to produce art works. Aesthetics were woven into the entire exploration process rather than being limited to the "artistic" features of the resulting products. Participants enjoyed creating art out of everyday materials and worked with their minds and hearts to contribute to the conceptual artistic design. Failure had no place in these activities and everybody could feel empowered to be an artist because the artefacts were considered not as the target of the activity, but rather as a means of participation through expression and of communication with others. A constructivist approach informed the work: what was important was not only the artefact itself but the actual learning processes that happened through interactions with content and concepts that were personally meaningful to individuals

(Martinez & Stager, 2013). The use of enquiry language and instigation questions was very important and facilitation was rather challenging for the artists, the educators and myself as it required appreciation of individual knowledge and support of free artistic creation. The arrangement of the working stations, accessibility, flow of participants and general management of the interactions all contributed to the creation of a context that encouraged doing and learning.

The projects, activities and events described later in this chapter are an integral part of communicating science at the Science Studio. As an important component of the outreach programme, they project within the community the reflections on content, pedagogy, art and science that happen within the studio, a participatory platform where artists and audiences are encouraged to become "active partners". It reasserts the importance of the community and the idea that the studio should be a place of dialogue and not a place of facts. In this frame, collaborative expressive art pieces made during activities and displayed in the centre become tools that stimulate ownership and affiliation (Simon, 2010).

Participation is related to democracy and freedom of choice (Bruyas & Riccio, 2012), and the empowering message in this type of activities is that everyone should be free to personally explore the world and express his/her individuality and that this can be done by relying flexibly on the skills and mind frames of both science and art.

A selection of tested activities: between art and science, learning and expression

If art and science were not to invade their specific territories, then activities could balance the characteristics of each. While designing the activities, there were always discussions on how much science content and how much art practicing should be presented. This is very similar to what happens whenever a science exhibit is designed or an exhibition is curated.

Sometimes, the scientific/technological contents and learning were more explicit and visible. In other activities it was art taking the lead and opening up a space for socio-scientific issues or individual self-expression. In general, art could rely on new tools and explore new issues, whilst science exploration gained tangible visual results that contributed to better learning as there was a push towards unrestricted expression and exploration that became meaningful for the individual.

In the activity *Curtains* – where mirror imaging was used to create prints on fabric – content knowledge took the lead in the process, yet art was in direct relation with learning: if you explored and learned how a mirror works, you could create more beautiful prints. In this activity children were engaged in exploring their abilities to flip images left to right and draw a "wrong sided" image on sanding paper that was later ironed on fabric resulting in a mirroring "correct print". In Saffa, near Ramallah, the drawing theme was stimulated by participants themselves. In fact, most of the children in the village knew how to draw Handala, a cartoon figure by the Palestinian artist Naji Al Ali. They used this as an important symbolic figure in their learning. In choosing to draw this known figure they overcame the distraction of having to imagine a new subject and they focused more on the scientific exploration. In Jerusalem, the children came up with a rainbow theme and started drawing what seemed to be more geometric patterns. Again, the drawing of the subject was not the centre of the activity: it was the visual phenomenon that was absorbing attention and encouraging exploration.

In the activity *Signing with Chlorophyll* and chromatography collage, art was used to stimulate a conversation about environmental issues and personal responsibilities. Participants first extracted chlorophyll, then signed their names in pencil on filter paper, circled them with colouring pens and added the chlorophyll extract starting from the centre. Chlorophyll separated into chlorophyll A and B so participants discovered that this substance is made of different components that have different shades of green. Then colours from the pens separated into their basic colour components and radiated beautifully while they diffused in the filter paper. Graphite instead did not diffuse, so names remained readable and fixed at the centre of the picture. These metaphorical representations stimulated conversations about the role and responsibility of humans within the natural environment. Chromatography was used both as a scientific way to separate chlorophyll to obtain beautiful pictures and to visualise a metaphor that triggered dialogue. One same technique encouraged both aesthetic, scientific and philosophical thinking.

In the *Blooming April Tree* activity, science was brought in through food colour mixing and experimentation with acid-base reactions using vinegar, baking soda and soap. The artistic and emotional stimulus came from the idea of reconstructing a traditional social context of the Old City of Ramallah. In fact, before urban design changed the area, a tree used to stand at the centre of the neighbourhood as a social meeting point. In the activity, people were invited to recreate that tree. The excitement resulting from this meaningful connection, drove participants in their exploration so that they were striving to generate beautiful combinations of colours which, in return, demanded deeper exploration and understanding of the phenomena. These were clear cycles of art and science learning fuelling each other in turn.

In *Bringing Back the Sea to Ramallah* the scientific focus was on palaeontology, a topic that does not generally call for great interactivity, as fossils are often displayed without encouraging dialogue among viewers. In this case, the activity was conducted in a public natural park where fossils were exhibited in their natural "habitat" and directly connected to their environment.

Real fossils were used to demonstrate how fossils form. Moulding and casting and an artistic activity with cyano-photography introduced the concept of a "physical shadow" of a once existing being. The mystery of exploring a shape that no longer exists opened up a space for imagination and spatial thinking as well as the need to explore the phenomenon more in depth in order to reach the understanding of something that once was and now is no more. This activity also encouraged an emotional connection to a social theme: in fact, with the territorial occupation, many Palestinian children from the West Bank have never seen the sea. This brought higher levels of motivation both in the exploration and in the self-expression phases of the activity. Participating children were allowed to draw the sea on the walls of the park. They drew the sea as they "know it today" on the walls of the sea that "used to be". Food colours and flour were used to make naturally degradable paint. We left the park with children's drawings of fish floating on the walls among the trees for the whole community to enjoy and appreciate.

The activity called *Life Back in a Shadow* started off as an exploration of light and shadows. Seventh grade girls from Jerusalem were invited to work in small groups and make a foil sculpture of a bride. They were then given torches to explore the shadows cast by their sculptures: they explored distance, angles, and different sides of the sculptures. Next, they outlined the shadows on canvas and they were asked to put tools aside and try to explore the shapes that they could recognise in these shadows. There came the surprise: in these shadows, the shapes of the bride sculptures could no longer be recognised. As they coloured them in different patterns, the girls began to elaborate the work metaphorically. The conversation evolved as they brought up ideas such as the fact that the future of a bride is initially unknown and unpredictable as is the shape of a shadow. Or that different brides have different futures which can be represented with different patterns. I must say it was not an easy activity to facilitate: in a social context, the seventh graders found out that they shared common fears and questions. When parrots and cats surfaced out of the abstract shadow shapes, the girls said that this could be connected to their longing for nature. It was amazing to explore the young girls' ideas and emotions and future experiments should be supported by research in pareidolia and other related fields of psychology. Here the initial "scientific" exploration of a natural phenomenon became the pretext for the exploration of self and others.

Interaction, iterations and engagement

The idea that interaction is a key component of learning is widely accepted. However, Edith Ackerman (1993) discussed interactivity from multiple perspectives. And I agree with her that interactivity is not limited to "hands on" but should also include "heads in". The "hands on" is not a set of procedures to be followed by a "heads in" of observations and conclusions. It is more of a combined "heads in on hands on" process. The hand makes a thought that is taking place in the head tangible. It is the pairing of knowledge and skill that constructs and expresses new learning.

In the "curtains" activity, children drew while imagining what would result from flipping the original figure in their mind. Sometimes they were satisfied with the results, others they would use a mirror to look into the reflection and draw accordingly. This meant that, although the concept was not yet being mastered by abstract thinking, the meaning and understanding of the phenomenon was being embedded in the cognitive structure of the child. Children were invited to use the actual tool (the mirror) instead of their imagination until they could do without it. One of the most interesting explorations was that of trying to write in the opposite direction. It was as if the colouring pens were stuttering. The children were going in cycles of mind revising handwriting. Each cycle was a stutter that showed the "heads in on hands on" process of learning.

When referring to interaction and engagement, the evaluation team of the Exploratorium (San Francisco) described cases of "prolonged active engagement" (Humphrey *et al.*, 2017). In their case the concept referred to science exhibits, yet I tend to think of art-science activities as living exhibits in progress. In both cases, when planning the design of an exhibit or activity, one should keep in mind various dimensions of the participants' experience: the physical dimension (what they are asked to do); the intellectual dimension (how they think, understand and conduct enquiry exploration); the social dimension (who is participating in terms of age, gender, specialisation, engagement, grouping); the emotional dimension (what feelings emerge, how satisfying is the experience, which relationships are established). Each activity will incorporate all these aspects. In the case of art-science activities, the physical dimension involves participants in dealing with science contents to then developed artistic products; for example, in the shapes of shadows. The intellectual dimension reveals itself in the processes with which participants explore scientific phenomena (such as light and shadows) and in the acts of design and expression that they perform when creating their own artistic product. The social dimension of the activities is particularly important, as emerges from the great attention dedicated by activity designers and facilitators to collaborative procedures, conversation, design of the workstations, and creation of collective art products. The emotional dimension – perhaps the most evidently connected to the specific nature of art – is fostered through the idea that all exploration must be meaningful to the individual. Emotions are also a key element to gain the attention of participants and increase the long-term memory of learned components and skills (Tyng *et al.*, 2017)

Conclusion

One of the most significant outcomes of the activities described above was that science teachers and artists engaged in planning and facilitating together, experimenting first-hand examples of art-science interaction and its impacts. By agreeing to participate. both crossed boundaries set by their disciplines and convictions. It was an empowering experience that could not be matched by any workshop or course. Indeed, one of the artists declared that she could not have conceived it if she had not experienced it in person.

The activities described in this chapter are first steps of an ongoing research that keeps developing and exploring new techniques for the design of more complete educational experiences. Yet the different interventions suggest that people enjoy contributing to the creation of an artwork that represents their understanding, their skills, or their opinions. It is their voice in a collaborative product that allows them – and us – to observe their ideas, understand and learn.

Artefacts were self-exploring devices that had the power to bring out meaningful associations and activate imagination, creativity and thinking. They were chances to become more aware of self, knowledge, understanding and literacy in science. Art became a celebration of acquired knowledge or a declaration of opinion; once participants understood how reactions worked, they began to create more colourful leaves; when they mastered drawing reverse images the rainbow designs grew; agreeing to sign with chlorophyll was a declaratory act; rebuilding a community tree was an act of resistance to urban changes.

Hands-on, interactive, participatory and collaborative science-art educational experiences inspire explorations that call upon both artistic and scientific skills within participants. Educators create these activities to explore content through interaction and experimentation with science and/or art techniques. In this context, art allows to celebrate multiple perspectives, open unexpected solutions and teach thinking through and within a material expression (verbal, visual or other). The added value of these activities is the creation of a situation specifically designed to foster expression, awareness, development of social values and exploration of the self.

Alongside the scientific content, art helps demonstrate individual understanding, reveals identities and highlights personal meaning about what is being learned: key features of any successful learning process.

Bibliography

Ackermann, E. (1993) Tools for Constructive Learning: Rethinking Interactivity, in: *Epistemology and Learning Group*, No. 15, Cambridge, MA: MIT Media Laboratory.

Bruyas, A.M. & Riccio, M. (Eds.) (2012) *Science Centres and Science Events: A Science Communication Handbook*, Milan and New York: Springer Science & Business Media.

Humphrey, T., Gutwill, J. & Exploratorium APE Team. (2017) *Fostering Active Prolonged Engagement*, New York: Taylor & Francis.
Martinez, S.L. & Stager, G. (2013) *Invent to Learn: Making, Tinkering, and Engineering in the Classroom*, Torrance, CA: Constructing modern knowledge press.
Simon, N. (2010) *The Participatory Museum*, Santa Cruz, CA: Museum 2.0.
Tyler, C.W. & Likova, L.T. (2012) The Role of the Visual Arts in Enhancing the Learning Process, *Frontiers in Human Neuroscience*, 6, 8.
Tyng, C.M., Amin, H.U., Saad, M.N. & Malik, A.S. (2017) The Influences of Emotion on Learning and Memory, *Frontiers in Psychology*, 8, 1454.

5.4 Answering the unasked questions

Ian Brunswick and Andrea Bandelli

Introduction

There is a seductive myth that Apple doesn't do market research. An earlier incarnation of this is the supposed Henry Ford quote, 'if I asked people what they wanted, they would have said faster horses'. Apple do in fact utilise market research and Henry Ford probably never said that. But the popularity of these anecdotes reveals what many people feel – that audiences don't always know what they want until they've seen it.

Focus groups, co-creation, co-curation and crowdsourced content sound great, and to many stakeholders they are the clear and simple calling cards of progressive public engagement with science.

The way that exhibition development works at Science Gallery Dublin is not as participatory and crowdsourced as people might think. Is this a contemptuous disregard for our audience? What are the limitations of participatory methods like focus groups and co-curation? And what are instead the benefits of taking a more ethnographic-inspired approach to designing exhibition experiences?

The Science Gallery process of exhibition development offers an opportunity to see how understanding and empathising with your audience may be as meaningful as seductively simple "participatory" processes.

To be clear – focus groups can be good ways to understand your audience's perceptions, beliefs and habits. They are less useful for understanding their needs, the environment of their daily experience, and the factors (social, political, economic) that shape their lived experience. Focus groups – which are principally a tool for understanding perception – are sometimes utilised as participatory decision-making committees. This is usually just a lazy outsourcing of research and creative development – 'Let's not think about our audience, let's just give them what they ask for.' It is a process that is vulnerable to availability bias; audiences are likely to ask for what they know. It is possible for this radical letting go of power to produce something that will challenge, surprise, activate or move audiences. Kiasma, a contemporary art gallery in Helsinki, famously crowdsourced the curation of a new exhibition following general dissatisfaction with the institution, titled *Make a Better One Yourself, Then!* This empowered and re-engaged an estranged audience, but it was a temporary and extreme measure in response to an immediate problem.

When we zoom into science engagement institutions, however, co-production and co-curation may actually reduce the diversity of the contributions that the public can make, and engage an audience that quickly stray from the perspectives it was originally consulted for (Bandelli & Konijn, 2015a). Asking the public to step into the content development process may simply reinforce the institutional culture of the organisation – the museum becomes an outlet for like-minded people. Why is that? One reason is that science engagement institutions still have a rather high-level "entry barrier": they tend to attract audiences which are already engaged with science and identify themselves with the institutions they visit (Dawson, 2014). Another reason is that public participation in science exhibitions is inextricably linked to the social and political impacts that visitors expect from public engagement. We're probably still far from answering James Wildson's question: 'If public engagement is a means to an end, what is that end?' (European Commission, 2008), but the public doesn't ignore the question when they are asked to participate (Bandelli & Konijn, 2015b). The extent to which the public is ready to contribute depends on leverage the museum can exert.

There are numerous exceptions where radical crowdsourcing was successful and even necessary, but it can be argued that empathy and observation of audiences is potentially more powerful than many direct forms of public participation.

Lit reviews and burger joints

Exhibition concepts arise in many different ways at Science Gallery Dublin, just like anywhere else. Some are generated internally, some come from our Leonardo group (a kind of think tank of 50 individuals from various fields), and some are proposed by outsiders in the research community, art world, or our audience indeed. Once an exhibition theme gets traction, however, it goes onto a kind of conveyor belt of processes that is not unlike researching a documentary or a magazine feature.

Many of the early processes are predictable: the programming staff read books on the topic, we meet with academics who specialise in the area to learn about their research, and we generally research the people, organisations, artists and artefacts that are relevant to a topic, even if the theme is initially a very broad concept. Usually it quickly becomes clear if a theme is too wide or too narrow, in which case we need to identify subtopics or related topics and decide which are most promising to pursue. An exhibition on conflict? Too broad; let's consider focusing on the phenomenon of longstanding grudges. An exhibition on deforestation? Quite specific; let's look at sharing at different scales and the tragedy of the commons. Once a theme gets moving on this treadmill, it isn't guaranteed to become an exhibition, it just means we are testing its viability and examining whether Science Gallery Dublin is well placed to address the theme in a new and meaningful way.

Where is the public in this early stage of the process? Well, sometimes we field-test exhibition themes in the guise of public events. It could be a panel discussion or debate, a workshop or even a performance – some taster of a topic that lets us debut a low-commitment piece of programming with our audience. We can see what the take-up is, what questions are asked, what concerns an audience raises, and the demographics of who shows up. Not terribly scientific, but like any prototyping process it is easy, fast, and repeatable, so it yields valuable informal knowledge about our audience's relationship to a topic.

The real participation, however, is already happening and just needs to be observed. It is naïve to think that as institutions trying to engage wider audiences with science and art we can create and control the best methods of participation. There are many extant theatres of engagement where our audiences engage with science and art every day, even if they do not know it, even if the interaction is problematic or negative, and even if the interaction is not rooted in the language of art or science. These environments of experience are real and ever-present, and as institutions we ignore them at our peril. To try and create an experience entirely independent of our audience's real-world experience is not only a fool's errand – it misses a massive opportunity to utilise familiar aesthetics, to subvert and alter familiar everyday interactions, and to get audiences to interact in ways that are familiar to their daily experience, but with a twist that may push them to adopt a critical lens without being pushed so hard they feel estranged or unwelcome.

Take, for example, Science Gallery Dublin's exhibition *FAT: It's delicious* (Science Gallery Dublin, 2014a), which looked at the science and culture around fat. This is a textbook example of appropriating an existing interaction and aesthetic – the atmosphere was that of a retro diner full of chrome and pastels, plastic french fries baskets and kitsch stand-up laminated menus. Audiences only needed a few visual cues to know their role was to sidle up to one of the red vinyl bar stools and place an order with the apron-wearing attendants behind the counter. Instead of burgers and chips, the menu had research experiments that the visitors could "order", and after signing a brief consent form, their number was up and another attendant (or "mediator" in science gallery language) took their BMI, measured their glucose levels, or ran them through cognitive tests. Other menu items were artworks, artefacts, or occasionally, actual food – a flight of oils that the visitors could taste to understand the different culinary properties of various fats.

If it seems a bit obvious to have a Jack Rabbit Slims–style exhibition space, consider the context. In early discussions about FAT, we encountered diametrically opposed public dialogues – that waning position that dietary fat was all bad, and the waxing notion that eating fat was actually good and obesity was all sugar's fault. More importantly, we were aware of negative and hurtful dialogues surrounding fat (particularly with regards to young adults) and the sometimes unhealthy obsession with (not) consuming fat. So we needed to address the topic in a positive, productive way, and also convey that our focus was more on fat as a substance than on obesity. In the end, we had a topic on our hands that people felt confused about – was fat good, was it bad, how much is right – and a topic that generally was discussed in a negative problem-oriented way. A diner offered the perfect, if extreme, counterpoint. Our exhibition team scouted local retro-themed diners, collected ephemera and observed customer interaction. In the end we got a bright, playful platform for a dialogue about a much-maligned but fascinating molecule.

A similar example on a very different topic was Science Gallery's *Sound Check* exhibition (Science Gallery Dublin, 2017a) which brought together music and the Maker movement. The idea developed from Science Gallery Dublin's involvement in Dublin Maker, a Maker Faire–like yearly event at which there were numerous makers, hackers and DIY enthusiasts often exploring music and sonic arts. The exhibition was going to combine two established cultural domains, so we engaged in some basic design ethnography (though never called it that) similar to the way we scouted diners for FAT.

In this case, staff documented two types of spaces in particular – hackerspaces and musical instrument shops. What was the entryway like, who hung out there, what did people do when they were there, was there signage and did newcomers get stares when they entered? And what objects and aesthetics were central to the two spaces' identities? It quickly emerged that the two spaces had different and possibly irreconcilable atmospheres, but we might be able to adapt the kinds of hands-on experiences that existed in both domains. We tried to replicate the opportunities to admire and test out instruments at guitar shops, while at Fab Labs and makerspaces there was an atmosphere of getting stuck in, of being present by being involved and making something, and we tried to replicate this interaction as well in *Sound Check*. The most useful part of analysing these cultural spaces, however, was identifying a major challenge: both domains had a history of being highly gendered spaces, and unwelcoming to the uninitiated (though this has been changing for some time, especially in maker communities, and is by no means a universal history). It was easy to imagine combining the two inspirations and ending up with a doubly 'male' space: a worst-of-both-worlds environment that felt like something between the AV room in *Stranger Things* (no girls allowed!) and the guitar-fanboy-fantasy basement of *Wayne's World*. Overcoming that potential barrier to entry was a key component of the design brief for the 'look and feel' and interaction design of *Sound Check*. It couldn't feel like a space unwelcoming hackerspace newbies, and it couldn't feel like a shop where you weren't welcome to pick up an instrument unless you were a guitar god.

Prototyping to surprise (no penguins allowed)

This emphasis on observation of audiences and environments doesn't mean we never involve our community in the design. We prototype exhibition experiences with our audience frequently, but it is a process for us to understand our audience better, not just consult them and expect that they can give as much time and insights in the curation process as professionals. Our goal is to develop something inspiring and challenging, but one in which audiences can naturally and nearly subconsciously find themselves becoming engrossed. It's a terrible sounding term, but we say that good exhibition design sometimes lets you 'zombie' into something before you realise how deep you've gotten. You enter a space, or are handed something, or you start playing a game before you have time to say 'what is this, do I like this, will I do something else instead?' Audiences who can sense what to do before they have to think about it are more likely to role-play, to ask meaningful questions, and engage in thoughtful critique because they are spending less time evaluating whether they should do it – they just do it.

For an exhibition on climate change, we identified that Science Gallery's target audience were inundated with messaging about the issue and were essentially at saturation point (Science Gallery Dublin, 2014b). We needed to present a different take on climate change or risk falling on deaf ears. Empathising with our 15- to 25-year-old audience was key – they were the target of constant messaging about reducing their carbon footprint, but also about the inevitability of climate change in the coming century. Add to that the difficulties of millennials getting jobs, mortgages and so forth, and our audience seemed to feel disempowered to enact the necessary large-scale changes that an older generation of leaders were refusing to make. In this scenario it's easy to tune out the doomsayers and believe that the problem is too big to solve through virtuous individual action.

In fact, we took that as a starting point, and flipped the narrative so that the exhibition didn't focus on individual action for collective gain, but rather focused on naked self-interest. The show adopted a (callous) internal-only tagline to sum up this approach: *Strange Weather: Fuck the Penguins, My Basement Is Flooding*. This flippant attitude freed us to speak to our audience in an entirely new way – about how they can be one of the 'winners' in a world wracked by climate change. If this seems crass, consider that as a starting point it forces an audience to accept the notion that there will be winners and losers as a result of climate change (ta-da! We've introduced the concept of climate justice the minute they've entered). This stark approach presents climate change differently than the bombardment our audience receives daily from pop culture, public messaging, corporate greenwashing, and environmentalists alike. In a word, it is a surprising approach, designed to jar with their everyday experience, hopefully getting them intrigued instead of dispirited and resigned like so much typical climate messaging.

The prime example was the *climate bureau*, a booth styled as a something between a career counselling session and a travel agent's office, developed for Science Gallery by artist-led think tank Co-Climate.[1] Again, we studied the sales pitch and the material culture of both settings and reproduced it so that audiences felt an uncanny familiarity – that they were taking a seat with a salesperson who was there to understand them but also sell them a choice, all in an aspirational, slightly neoliberal atmosphere. The bureau attendant guided visitors through 10–20 year career choices that would capitalise on melting permafrost, algal blooms, and other natural phenomena with a 'you, yes you, can get rich too' air. (We considered a commission system for mediators, but decided that was maybe going a bit too far). Audiences left with a prescription of skills, languages, and geographic relocation they should undertake, tailored to their own talents and interests that would also put them in a position to avoid the worst effects of climate change and capitalise on its opportunities. Cynical, perhaps. But it was eerily believable, and left visitors woke to the fact that the effects of climate change would be felt worst by poorer populations unable to rapidly adapt, and therefore reinforce growing trends in injustice and inequality.

No simple answers on how to keep it simple

All these examples so far have focused on a highly interactive element of Science Gallery programmes, and these elements can sometimes be the parts of an exhibition with the highest barrier to entry. Getting that barrier to entry as low as possible is where prototyping with an audience can be particularly fruitful. During the early development of *Field Test* (Science Gallery Dublin, 2016), an exhibition about food and agricultural futures, we experimented with giving visiting students tin cans with blank labels and letting them imagine strange foods. What if they had to design the most disgusting food they could imagine? The most healthy? The rarest, tastiest, or most sustainable? We evolved the prompt based on the results we got, and though small, it eventually became an integral in-gallery part of the exhibition where visitors could propose and share ideas for new and innovative foods.

For *In Case of Emergency* (Science Gallery Dublin, 2017b), we wanted to get strangers talking to each other about the end of the world in the same way they might discuss the weather at a bus stop. But trialling a highly-interactive element showed that strangers didn't loosen up quickly and were quite guarded. So we inserted a short icebreaker card game beforehand, where each card was a solution (often preposterous) to a global catastrophic risk. With limited choices, each participant had to advocate for a course of action that might not be their ideal choice, but it was the card they were dealt. Loosely based on the way a game of Apples to Apples or Cards Against Humanity can get out of hand even with the most civilised participants, it did the trick. This bit of role-play meant they jumped in head first and in no time were loudly advocating for asteroid missile defence or global chocolate rationing, and suddenly the ice was broken. The deeper, more serious discussion of real global risks proceeded much more successfully.

Audience feedback can unintentionally create barriers, too. We sometimes find that what audiences suggest and what they actually like are subtly and directly at odds. For example, for a show on failure we discussed what failing meant to different audiences – researchers, university undergrads, young entrepreneurs, job seekers, secondary school students, parents, artistsand so forth. Many groups said that they wanted people to be more open and accepting about failure – that we should create a space where people could share their greatest failures and "get them off their chest". We did a Post-Its and pens test of this and it stumped people. It's a deceptively big question, and before people could put pen to paper, they were soul-searching and weighing up pretty heavy life experiences. In fact, it was a downright paralysing question, and any answer that fit on a 10 cm piece of paper seemed trite. We tested different phrasings of this question again and again, but the problem seemed to be that most questions had so many possible answers that it forced people to review their entire life experience. Then we hit on "tell us the last time you messed up". People immediately had *one* answer. Sometimes humorous, sometimes tragic – it yielded so many different answers. In fact, they often ignored the prompt and answered with the failure they really wanted to talk about. The key is, it wasn't the question that our audience asked us to ask, but it was the one that got them discussing what they wanted to discuss.

Pop culture and polarised debates: an opportunity for art and science

Lastly, some of the most important issues at the intersection of art and science are contemporary and rapidly evolving. They are being discussed and debated in multiple fora and audiences may be experiencing simultaneous conflicting narratives, and this is where empathising and understanding our audience's experience has been essential. In scenarios where Science Gallery produces an exhibition that taps into a cultural zeitgeist, it has been essential to survey the pop culture and communication landscape shaping our audience's experience of the topic. Two recent examples include *In Case Of Emergency* (ICOE) and *Humans Need Not Apply* (HNNA).

For *In Case Of Emergency*, we focused on three areas:

1. Apocalyptic and post-apocalyptic cultural texts (everything from *The Walking Dead* to Thomas Cole's *Course of Empire* series, etc.).
2. The Prepper movement and its variations (from off-the-grid rural preppers to Silicon Valley and Ireland – the Science Gallery staff even did a survivalist away day with some ex-military trainers).
3. Academic fields focussed on assessing Global Catastrophic Risk (particularly the work of Oxford's Future of Humanity Institute).

All three research directions were valuable and helped us realise that our 15- to 25-year-old audience was quite familiar with pop-culture portrayals of catastrophic risks in graphic novels, cinema, and television, but had little awareness of Prepper subculture. This meant we could use visual cues and physical culture that overlapped the two, but we couldn't assume familiarity with the concept of a Bug Out Bag or why having a house in New Zealand is so attractive to uber-rich preppers. And perhaps most interestingly, while we found that young visitors were unsurprised when we outlined the major risks identified by researchers like Nick Bostrom, we also found that scientists working in fields relevant to those risks had significant concerns about public engagement on the topic of catastrophic risks. Some felt that public discussion exaggerated the risk, or in some cases they felt that their research would be portrayed as culpable or contributing to potential catastrophes. These cultural surveys helped us realise that our audience was more ready and willing to talk about the difficult questions than many scientists, but also that we could assume universal familiarity with the pop-culture of apocalyptic portrayals but not the sub-culture of apocalyptic preparedness.

Second, when *Humans need not apply* (Science Gallery Dublin, 2017c) was being developed, articles about the promise of neural networks and machine learning were frequent but matched by equally numerous articles on the impending economic apocalypse from robot-induced mass redundancies. We embraced this debate and we mocked up a visual identity of the show that featured lots of the positive AI narratives with things like Google's Firefly car (imagine a happy koala emoji with wheels), Baxter the helpful-looking 'General Purpose' robot, and other life-improving concepts, designs and innovations. Alongside that, we developed what you might call an 'evil twin' set of posters and moodboards, filled with things like Boston Dynamics' surprisingly terrifying SpotMini robot (it washes the dishes with its face), and artworks that explored anxieties around automation decimating human employment. Creating these exhibition mockups encapsulated the polarised public dialogues on AI and automation, and let us gauge people's reactions to the two different directions. Which one matched their feelings about robots? Which one did they want to see? Of course, any in-depth treatment of the topic would include a bit of both, but we needed to address the extremist points of view to empathise with our audience's position: blasted by an onslaught of the two narratives, and probably having chosen sides. The resulting exhibition was uncanny but playful – trying for an uncomfortable middle ground somewhere between Siri and Skynet. The experience focused less on whether robots would take our jobs, and more on what it meant to be human or machine, and whether we could find meaning and value in creative works created by algorithms, robots, and neural networks. That angle, according to all our feedback and evaluation, surprised and inspired our audience, precisely because it bucked the reductive, binary status quo debate.

Surprising empathy

Exhibitions offer a unique way to explore art and science. They create a platform and a context to have those 'calm, constructive and even uncomfortable conversations on the kind of future we want' (Schwab, 2018) among multiple stakeholders reconciling opposing views. Through art we develop cognitive and emotional tools to imagine the future and achieve creative breakthroughs; we see differences not as threats but as new frontiers of human connections, which helps to build empathy. By preparing ourselves for possible futures, we also become more resilient and learn to absorb the shocks of the impossible and think of it as real. We learn to question the foundations of our own world views.

In the pursuit of engagement, our professional field may have overemphasised the simple-sounding promise of participation, and under-emphasised the power of empathy and ethnographic methods. This is not a rallying cry to revert to the attitude of "we know best". No one understands our audience better than our audience. To simply give our audiences what they ask for would absolve us of the more difficult task of walking a mile in their shoes and trying to understand their experience and environment. Furthermore, it may contribute to giving audiences familiar and unchallenging narratives. From its foundation in 2008, Science Gallery has challenged participating scientists, artists, and publics to think differently about what engagement means to them, and to empathise and collaborate across academic disciplines and social divides (Brunswick, 2017). This direction has met resistance at times, but it has yielded a unique and widely respected institution.

This chapter focused primarily on public audiences, but the same goes for the people whose work is exhibited at Science Gallery. Artists and scientists (and designers and engineers, etc.) are typically accustomed to different professional practices, and have different ideas about interacting with the public and across the art-science boundary. Just as with the public, we try to challenge the expectations of artists and scientists, who sometimes envision a more didactic platform and are surprised when we push them to think of the gallery as a two-way dialogue that can contribute to their research, influence their practice, and be more than an instance of 'selfless' public outreach. Not something they always ask for, but once experienced, is near-universally embraced.

For museums, it's easy to ask, but it's harder to listen. And yet, both are needed. If we observe, study, and listen to the whole experience of our audiences, maybe we can produce something even better than what they would have asked for, and in doing so, help everyone understand themselves that little bit more.

Note

1 For more see www.coclimate.com (accessed 15 July 2019).

Bibliography

Bandelli, A. & Konijn, E.A. (2015a) Public Participation and Scientific Citizenship in the Science Museum in London: Visitors' Perceptions of the Museum as a Broker, *Visitor Studies*, 18(2), 131–149.

Bandelli, A. & Konijn, E.A. (2015b) Museums as Brokers of Participation: How Visitors View the Emerging Role of European Science Centres and Museums in Policy, *Science Museum Group Journal*, 3(3).

Brunswick, I. (2017) Genesis of the Science Gallery, in: Trench, B., Murphy, P. & Fahy, D. (Eds.) *Little Country, Big Talk: Science Communication in Ireland*, 159–175, London: Pantaneto Press [Online]. Available at: www.weforum.org/agenda/2018/02/dont-underestimate-the-value-of-dialogue (Accessed 19 July 2018).

Dawson, E. (2014) "Not Designed for Us": How Science Museums and Science Centers Socially Exclude Low-Income, Minority Ethnic Groups, *Science Education*, 98(6), 981–1008.

European Commission. (2008) *Public Engagement in Science*, Luxembourg: Office for Official Publications of the European Communities.

Schwab, K. (2018) *Don't Underestimate the Value of Dialogue, World Economic Forum* [Online]. Available at: www.weforum.org/agenda/2018/02/dont-underestimate-the-value-of-dialogue (Accessed 19 July 2018).

Science Gallery Dublin. (2017a) *SOUND CHECK at Science Gallery Dublin* [Online video]. Available at: www.youtube.com/watch?v=a3nB4RwCjBU (Accessed 19 July 2018).

Science Gallery Dublin. (2017b) *IN CASE OF EMERGENCY at Science Gallery Dublin* [Online video]. Available at: www.youtube.com/watch?v=HL0vKqsGLTI (Accessed 19 July 2018).

Science Gallery Dublin. (2017c) *HUMANS NEED NOT APPLY at Science Gallery Dublin* [Online video]. Available at: www.youtube.com/watch?v=6y94BdW1LPs (Accessed 19 July 2018).

Science Gallery Dublin. (2016) *FIELD TEST: Explore the Exhibition at Science Gallery Dublin* [Online video]. Available at: www.youtube.com/watch?v=nQ8u8rf6_JM (Accessed 19 July 2018).

Science Gallery Dublin. (2014a) *FAT: IT'S DELICIOUS at Science Gallery* [Online video]. Available at: www.youtube.com/watch?v=HMTg0b94HRU (Accessed 19 July 2018).

Science Gallery Dublin. (2014b) *STRANGE WEATHER at Science Gallery* [Online video]. Available at: www.youtube.com/watch?v=v-DOryiV1s0 (Accessed 19 July 2018).

5.5 Case studies

5.5.1 The value of artefacts in participatory art

The creation of an artefact as a result of an educational activity may be of deep significance in the learning process itself. It is in fact a sort of "embodiment" of the experience, which retains personal meaning and a communicative capacity. Tangible artefacts with an added aesthetic value may come to represent something durable and valuable not only for the individual, but also for a community that is receptive to, or shares, the individual's political, ethical, cultural, or psychological positions.

Case study: A.M. Qattan Foundation – Science Studio, Ramallah (Palestine)

Samar Darwish Kirresh, Principal Researcher

AMQF is an independent, not-for-profit organisation working in culture and education in Palestine and the Arab World. The Science Studio project is working under ERDP in cooperation with Ramallah Municipality to lay foundation for a science centre in Palestine.

Bringing the Sea back to Ramallah

A science and art educational activity for children, inspired by Palestinians' nostalgia of the sea, of which they were deprived by the Occupation. Here art is used not as an entry point to discuss a scientific topic, but as the culmination of a learning process. The mix of art and science based activities encourages learning processes that fuse facts and imagination, knowledge and feelings.

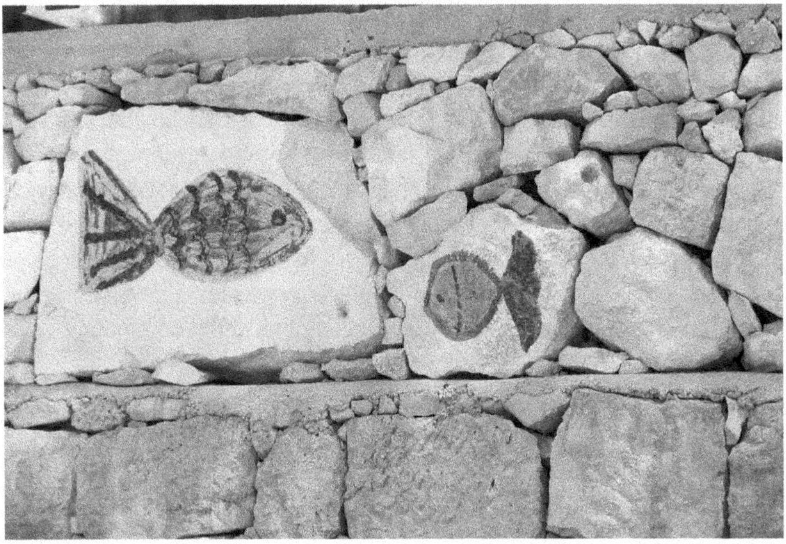

Figure 5.5.1 Collective work by students who drew the sea on the walls of a public space as a result of a learning art and science program

Source: A.M. Qattan Foundation

Description

In 2016, A.M. Qattan Foundation participated in the Nowar Nisan Festival, organised by the Ramallah Municipality, with educational activities for children focused on "The Sea".

On this occasion, the Municipality supported the transfer to the Al Radaneh public park in Ramallah of four teachers and 40 students with hearing disabilities aged 10 to 14.

Participants were involved in a five-hour long activity that brought together elements of scientific knowledge associated with palaeontology and artistic expression connected to painting. The workshop opened with a shared discussion about continental drift and marine life from millions of years ago. The conversation was moved forward by aid of enquiry questions explored through a laminated picture book designed especially for the project.

The children then had the opportunity to work with a group of young students engaged in another fossil project who shared their practical work on fossil samples entailing collection, classification and acid tests.

Following these strictly scientific introductions, adult volunteers and a group of science teachers facilitated other activities such as the production of cyanotype photographs, fossil hunting, moulding and casting in sand.

Activities were wrapped up by an open discussion on the sea and on the emotions surrounding the fact that it had been made inaccessible by Occupation with a sudden realisation that all Palestine was once "The Sea". Emerging feelings converged in the final artistic production, as children were invited to paint their own fish on the walls of the park. Leaving a public mark of the work done, which was lovingly accepted by park staff and appreciated and embraced by the whole community.

Outcomes

Nostalgia for the sea was the real thematic heart of this activity, but science was seen as an immediate and natural connection to the theme, that allowed to give concrete meaning to the activity. The initial phase of providing facts about ancient marine life, the following activities and the final process of freely drawing the sea, inspired learning cycles that involved imagination and construction of knowledge.

A frame of scientific and art experiences was necessary to seize scientific concepts and satisfy a sense of belonging to the sea.

Specifically, the art-based activities were acts of communication. The aesthetic outcomes ultimately stood in a public communal space, also empowering children against misconception on their disability. The possibility of working in public spaces indicated the municipality's trust, confirmed by institutional posts on Facebook and websites. The community responded passionately, as park visitors kept asking about the drawings until they faded away, just like the Sea.

The activity was seen as professionally empowering for the adults involved, who described as a profound experience the way they saw an apparently random group-painting evolving into one powerful collective image. The free, non-restricted context created opportunities of both expression and science learning. One researcher said: 'I had not touched colours for long. For a grown up like me it was fun. The children were helping me. They were the masters. They did not feel deprived'.

Partnering with teachers was very empowering and contributed to the integration of art-science learning in both formal and informal learning environments.

Art applied to science workshops gives the participants a new degree of freedom where self-expression becomes an important and integral part of the learning processes based on play and personal meaning making.

5.5.2 Open-ended exploration

Science centres and museums have a long tradition of study and practice in informal education and lifelong learning. Studies in this field follow different strands of research that range from the application of problem-solving methodologies to craft making.

So-called tinkering activities are based on the idea that in order to truly understand something, it is necessary to let one's own learning process emerge during an activity that is personally relevant. In order to achieve this, tinkering-like practices have paired the technical act of building with a way of operating that is typical of artistic research. In what sense? Usually, scientific research and technical practice require detailed planning so that the stages of work are mostly defined before starting the testing phases, often corroborating or denying a pre-formulated thesis. On the contrary, artistic research is frequently characterised by less predetermined processes. Individuals proceed through trial and error, adjusting their research and constructing paths in ways that are initially unforeseeable. In artistic research both the outcomes and part of the trail that leads to them are unknown. This specific way of progressing has a deep learning value, as it helps individuals to find their own learning style and method, becoming aware of one's personal way of knowing and producing knowledge.

Case study: The Tinkering Studio, Exploratorium, San Francisco, CA (USA)

Mike Petrich and Karen Wilkinson, Co-founders

The Exploratorium is a public learning laboratory that develops experiences for exploring the world through science, art and human perception. Founded in 1969 by physicist and educator Frank Oppenheimer, the Exploratorium envisions a world where people think for themselves and can confidently ask questions, question answers and understand the world around them.

The Tinkering Studio

An exhibit gallery and drop-in workshop for visitors of all ages to participate in making, tinkering and playful enquiry. The Tinkering Studio engages learners in blended science, art and technology explorations through aesthetically rich science investigations and collaborative constructions to ultimately empower participants and encourage autonomous critical thinking.

Description

Since 2007, Tinkering Studio designers and facilitators support inspiring, personally meaningful explorations and experiments with light and colour, sound and music, circuits and electronics, motion and mechanisms, and other phenomena-rich investigations. It is a chance for museum visitors to slow down and become deeply engaged in making something for themselves and with each other.

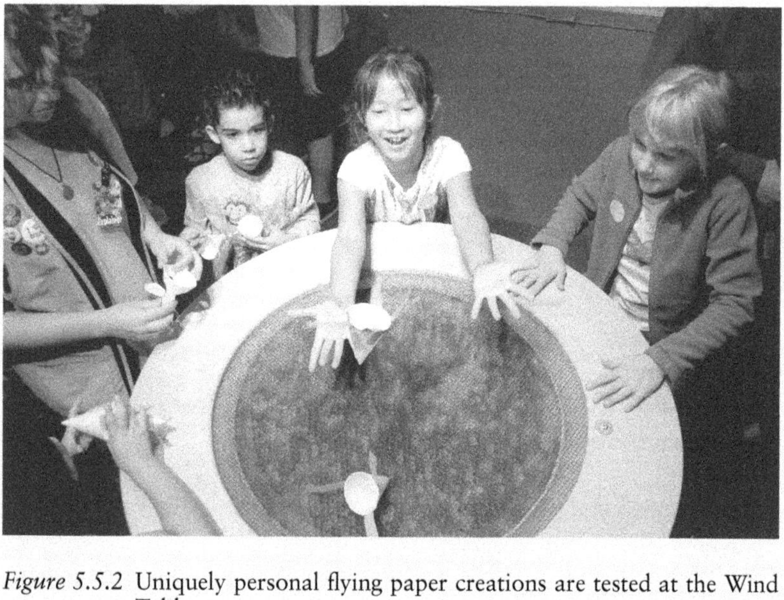

Figure 5.5.2 Uniquely personal flying paper creations are tested at the Wind Table

Source: Exploratorium

The Tinkering Studio supports learners to pursue their own ideas, along their own pathways of understanding, often revealing universally relevant design and facilitation strategies that are ultimately shared with a network of international partner organisations developing their making and tinkering approach to education. International artists, scientists and educators are all involved in the development and testing of new tinkering explorations with museum visitors.

The studio hosts residencies for artists and scientists to work together, co-developing imaginative new enquiries for the public. It also offers teacher internships aimed at supporting the facilitation of these new experiences, as well as developing the translation of the activities into formal classrooms or after-school settings. All activities and explorations developed in the Tinkering Studio are available as online "open-source" guides for educators to use in formal and informal settings.

Outcomes

Enquiry is at the heart of thinking and learning. The maker movement has been popularised in education in important ways. Educators and stakeholders are paying attention to the benefits of these kinds of construction-based learning moments. Informal educational organisations have much to offer in these conversations as many have been engaged in exploring science enquiry for many years. The Tinkering Studio blended science enquiry with the maker movement. Tinkering, which combines play and enquiry, is informed by the exploration of science phenomena at the Exploratorium and by the maker education programs proliferating around the globe.

Core to the pedagogical approach of the Tinkering Studio is a careful blending of science and art. Artists and scientists look at the world through a unique lens. Neither want to be told the "right" answer before developing their own thoughts and ideas about the world, and both are consummate questioners about the status quo. The expression of current thinking and ideas ultimately results in new genres of scientific research, and important new expressions in the world of art. The process of thinking for yourself and pushing one's own understanding to the limit is a trait that we highly respect in the visiting artists and design colleagues that we work with while developing new interactions for tinkering and making.

A tinkering disposition is the main outcome of the designed and facilitated activities, ultimately supporting a new generation of learners to think for themselves and ask their own questions about the world around them.

The meaning-making practices of scientists and artists are key to informing a constructivist approach to learning. Making careful observations of natural phenomena, exploring and investigating it with a variety of tools, technologies and materials, and working socially with others are key goals for supporting personally meaningful thinking and learning.

5.5.3 Participatory performances

In the 1970s, Augusto Boal's Theatre of the Oppressed refined some of the early theatrical experiments of the 1950s in which performers interacted with the public through dialogue.

Today, audience participation is quite common in theatrical performances and actors and directors have developed various ways of shattering the "fourth wall". In these experiences the shortening of the distance between performers and viewers can have tragic, uncomfortable, comical or engaging effects.

Numerous contemporary experiences connect actors and audiences through the surging of strong collective tragic emotions, while theatre forms such as British pantomimes and traditional cabaret require viewers to participate through laughter, cheering or insults. These conventions have been taken up by science show professionals that often integrate comical or surprising elements in their acts while encouraging an open dialogue with audiences that forces changes in every presentation, calling for skillful improvisation.

Case study: Hiša eksperimentov, Ljubljana (Slovenia)

Miha Kos, CEO

Hiša eksperimentov strives to awaken curiosity, stimulate creativity and impart critical thinking skills through open communication, exploration and discovery. It is a science centre that plays an active role in programming and science festivals within the city.

Science Adventures

A performance in which an actor plays out a show based on a general plot which adapts to the audience responses, through the art of improvisation, like in a jazz jam session. Performing arts have the ability to bring to life subjects through a direct interaction with the audience, prompting dialogue and actively promoting a culture based on questioning and enquiry.

Description

Science Adventures are being delivered at Hiša eksperimentov since 1996. They are Science Shows designed to trigger discussion about a specific scientific topic (such as water, motion, bubbles, etc.). A science adventure is composed of experiments and demonstrations which are connected into a story. The performer uses the Science Adventure as a tool to raise curiosity and stimulate discussion.

A Science Adventure lasts between 30 and 45 minutes depending on the interaction. It is performed in an auditorium or similar environment. Participating visitors can be groups of children, teenagers, students, teachers, elderly people, families, couples and more. Performers must be able to adapt the Science Adventure to the specific group of visitors they are engaging.

An author constructs the Science Adventure show (builds the story line, collects the demos and adds the humour) and runs, improvises and improves the Science Adventure as first performer.

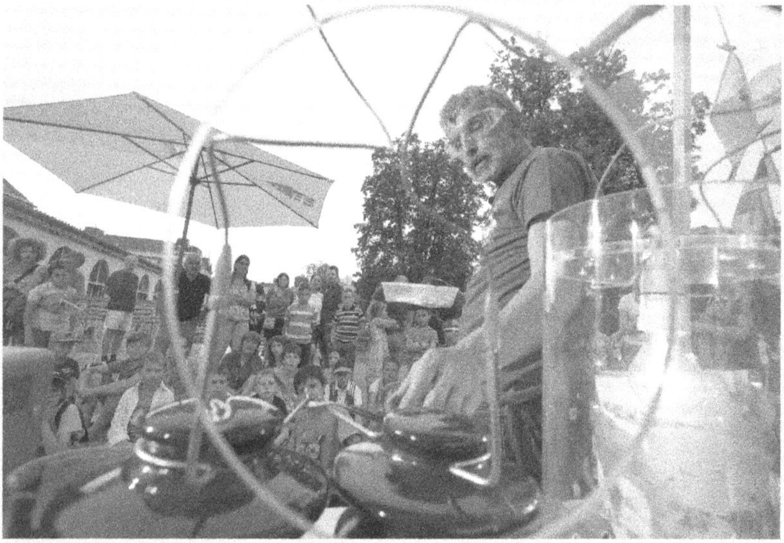

Figure 5.5.3 People participating in the science adventure Electrology

Source: Branko Čeak for Hiša eksperimentov

New performers are then trained to run the show. Each performer adds new values through experience, discussions and humour and work with the author of the Science Adventure to include flexible variations. These performances are living and evolving science communication tools. They are subject to constant adaptations and improvements.

The performer responds to the public, keeps eye contact with the audience and is constantly looking out for emerging curiosity and interests. Live questions and suggestions by the audience are promptly rewarded publicly and given the highest priority and importance. In fact, the point is not to complete the show but that of performing with and for the public, mastering the art of narrative and involvement.

Outcomes

Science Adventures are based on the art of narration and live performance, mixed with a healthy amount of humour.

The main aim is to empower participants to think critically, ask questions and seek answers by pronouncing the magical words: "I don't know".

If we look at the "hands-on" exhibits as the bones of a Science Centre, then Science Adventures are its meat. Despite the same set of exhibits, visitors keep on returning to the centre because of Science Adventures. It is like witnessing different improvisations of the same tune. Just like jazzing.

Science Adventures are video recorded to promote self-evaluation of performers and they have stimulated in time several other activities such as a yearly *Sciencetival* – Festival of Science Shows – and *Inspire* – an open school for making and running Science Shows.

Improvised and open-ended performances intrinsically value the process of questioning, as it is through questions that the audience interacts and becomes part of the performance itself. This ultimately results in the nurturing of critical thinking that is essential to the development of a critical mind-set. Humour also plays an important role in the performance, as it activates the brain and improves retention and long-term memory.

5.5.4 Learning from community centres

In the last decades, science centres and museums have been questioning and reviewing the specific role that they play within society. Who are they relevant for? How can they better answer the needs of their existing and potential audiences? In the attempt to reach out and respond to their surrounding communities many institutions have begun to include in their programming activities that cater for specific local groups, inspired by the practice of community centres. In this scenario, artists have often been brought into the loop to encourage participation of different local communities, connecting more freely with interests and inputs from the territory.

This effort to reach out to different individuals is often repaid not only by the engagement of new audiences but also by the discovery of local experts, amateurs and "champions" that become precious allays of the institution. Such community-oriented programs have even deeper effects when members of these audiences become consultants or even regular members of more formalised advisory boards.

Case study: MUSE – Museo delle Scienze, Trento (Italy)

Samuela Caliari, Head of Audience Development
Lucia Martinelli, Senior Researcher

MUSE is located, since 2013, in a beautiful building designed by architect Renzo Piano. It aims to interpret nature with the eyes, the tools and the questions of scientific research, seizing the challenges of the present and inspiring scientific curiosity and the pleasure of knowing.

Voices in the Square

In the hearth of the museum, during a figurative art exhibition representing urban squares, artists, scientists and citizens of all ages could stage a public message, a memory, a passion about various topic concerning science and art. The artistic setting and the invitation to use different expressive media encouraged a more varied participation, highlighting the intention of the museum to serve as an *agorà* for open discussion in the hands of its users.

Description

In 2017, the temporary art exhibition *La Piazza* (*The Square*) by local painter Matteo Boato served as setting for the staging of a series of 30-minute selected presentations addressed to museum visitors and online audiences. Just before the closing time of the museum, participants engaged in performances based on talks, dance or music with the simple aid of a microphone, a lectern and a chair.

Among 39 proposals (12 artistic performances, 8 scientific talks and 19 exhibitions blending art and science), 27 were selected. They engaged with several scientific topics such as zoology, limnology, sustainability, architecture and landscape. Presentations were selected by an interdisciplinary commission involving experts from MUSE and other affiliations. Finalists included 55 people, 17 male and 15 female, with an average age of 35 to 45.

The selection was based on four evaluation principles: scientific value; artistic value; communication style; and speaking skills.

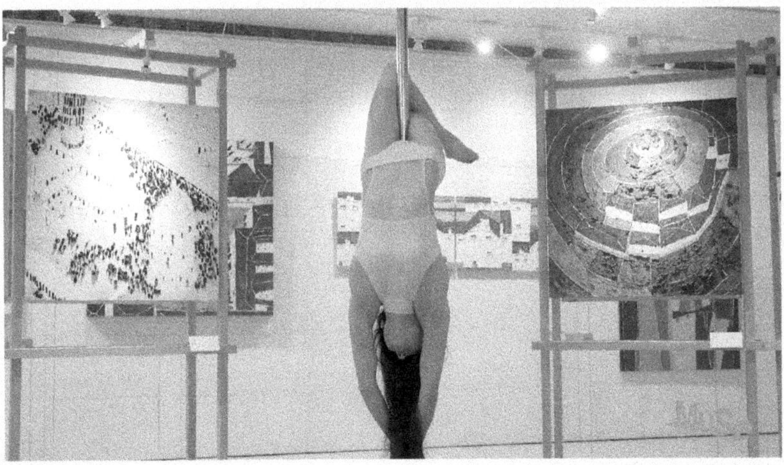

Figure 5.5.4 Pole Dance Soul Expression at MUSE

Source: Archive Muse

All selected performances were recorded and broadcasted live on MUSE's Facebook page, as one of the major goals of the project was that of reaching out to virtual audiences and communities through social media.

Outcomes

By involving citizens – not necessarily experts and/or professionals – with various scientific and artistic backgrounds, this project aimed at adding the socio-cultural dimension to the artistic and scientific disciplinary crossover, to connect the community and the territory in a collective expression and become an original form of social narrative based on art and science.

The rationale behind this experimental project was based on the following frame of mind: (1) museums of new conception are *agorà* of multidisciplinary expression; (2) passion is a feeling that scientists, artists and citizens all share; and (3) scientific engagement is enhanced by the capability to leave our own "comfort zones" and open our minds to new knowledge and information.

By assuming its role as a dialectic square open to all, MUSE aimed to intermingle scientific dissemination with creative sharing and citizens' involvement. In this regards, the artistic exhibition *La Piazza* (*The Square*) was valued as a suitable metaphoric setting for the hosting of different forms of expression and contents. The project reinforced MUSE's role as a space devoted to cultural and human aggregation where people's emotions and artistic expression can safely meet. It was rewarded by participation from online followers of the project, which reinforced the museum's identity, widened audiences and favoured the birth of a new network of contacts that have since then nourished various other projects at MUSE.

Opening up to new and mixed contents and forms allows for the engagement of wider communities, making space for the personal expression of non-experts.

5.5.5 Video-making and new content producers

Cultural institutions in general cannot ignore today's distributed availability of tools for producing communication and engagement products.

Social networks represent an extremely powerful infrastructure through which products, such as videos and images produced by individuals and groups, are broadly shared. Not all these products are of high quality, both in content and form. Creating occasions in which users can work with professionals in the field of content, such as scientific advisors, and in that of form, such as professional video makers and artists, can become an insightful way to contribute to the upbringing of more qualified content producers.

Working to develop critical and creative skills can help raise a general awareness in society of the importance and responsibility that we all have when we share information through social networks or more formal channels such as television or radio programs.

Case study: National Science Museum, Pathumthani (Thailand)

Ganigar Chen, Vice President

The National Science Museum is a government organisation focusing on presenting relevant science-based experiences for people of all ages. It is a group of science-based museums with the aim to improve the quality of individual learning and contribute to the development of the nation.

Science Behind the Scenes

An annual short-film competition open to students and the general public. The program invites people who are interested in science and film production to creatively combine their interests. The initiative is based on the general idea that this project can help form and support a new generation of science communicators who can effectively bring awareness of science to Thai society at large.

Description

Every year NSM and Thai PBS host the short science film competition program *SBC – Science Behind the Scenes* that connects filmmaking with science. The initiative is promoted as a competition which offers free registration for three categories: high school students, university students, and general participants. Teams from each category can submit short films through proposals and "trailers". Each year, the theme proposed by the competition committee focuses on a different science topic.

After submission, the proposals are examined by three groups of experts: scientists, who verify the content which participants want to convey and ensure that the presented concepts and data are accurate and up to date; film producers/directors, who determine the artistic quality of the videos, including creativity and entertainment factors that sum up to the participants' production abilities; science communicators, who evaluate whether the final products are appropriate for target audiences and balance the weight of arts and science.

Ten teams from each category of proponents are invited to join production workshops. Eventually, one winner from each group will be chosen by the panel of committees to receive winning awards and have the film broadcasted on the public television channel. Winning works are

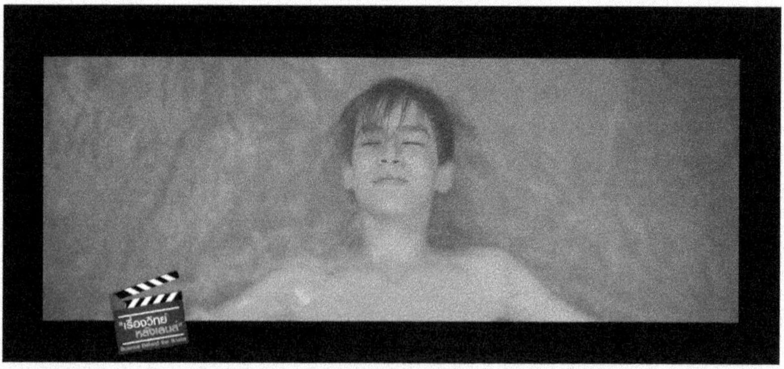

Figure 5.5.5 A scene of a short film of the Science behind the Scenes, broadcast on ThaiPBS public television

selected based on how they manage to represent science in ways that suit audience preferences and stimulate public interest in science.

The entire program takes around three months to complete. In 2017, nearly 100 teams of participants submitted their proposal in the first round.

Outcomes

The key message of this program is "science is not always boring". There are many artful and non-intimidating ways to communicate and help audience understand difficult scientific content. Such positive attitude and awareness that science is in nature and that it represents the absolute foundation of society can ultimately help create future generations of global citizens.

The program encourages the training of new science communicators, since participants in the program have a fair chance to train systematic thinking and reflect on legitimate science communication.

Moreover, helping media creators who have a limited scientific background to gain confidence and speak up for science could also encourage them to no longer alienate science in their future work.

The momentarily rare collaboration between prominent artists (famous Thai film producers and directors) and scientists would certainly strengthen the programme and help develop more interesting content as well as reach out to more diverse groups of audience whilst building confidence in the fact that science knowledge is not the inaccessible realm of scientists.

Documentaries are by nature a hybrid product, as they are artistic expressions in themselves. Yet they also have a strong intent to deliver specific messages. The mastering of these visualisation tools by lay people allows the latter to find and explore their personal narratives, often based on the language of the peers and thus more effective from the point of view of communication.

5.5.6 Curation and open calls

With the advent of digital communication, the work of curators has changed in many ways. One aspect of this transformation concerns the way artworks are gathered in collective events such as fairs, collaborative exhibitions and so forth.

Curators can now rely on powerful tools to broadcast open calls that will reach vast communities of emerging artists and creatives that might not be already known in the art market. In this scenario the curator, alongside programming teams and other museum professionals, plays a key role not only in the final selection process of artworks but also in the definition of the selection criteria, which become public and must be set beforehand.

Standards are to be identified by each individual institution and curator according to their particular nature and aims, yet science engagement institutions that have experience with open calls – such as Science Gallery, Ars Electronica, Ontario Science Center – frequently declare that they are looking for artworks that are: transdisciplinary; provocative and unconventional; open in format; with a connection to science and technology. Often a focus on an active role of visitors is also stated in the search for works that spark dialogue or facilitate interaction, collaboration and co-creation.

Case study: Ontario Science Center, Toronto (Canada)

Ana Klasnja, TechArt Fair Creative Director

An iconic Canadian cultural attraction, the Ontario Science Centre has welcomed more than 52 million visitors since 1969. It conceives and builds world-class exhibitions and serves Ontario education and innovation ecosystems to inspire passion for the human adventure of discovery.

TechArt Fair

A juried festival for tech-focused artists to present provocative projects, reveal their creative process, inspire excitement and spark interaction, collaboration and co-creation. This art-centred initiative raised visitors' awareness of new technologies, while opening the centre to new creative processes moulded by external creators and broadening its partnership network.

Description

In 2018, the TechArt Fair featured 20 projects, ranging from immersive virtual reality and computer-generated artwork to ceramic jewellery inspired by bioengineering, wooden automata, and a sound-based work transforming the orbital frequencies of the TRAPPIST-1 planets into harmonies visitors could play. The festival was based on the notion that technology has enabled the creation of art from the beginning of human expression and that 21st-century artists are incorporating new technologies into their practices.

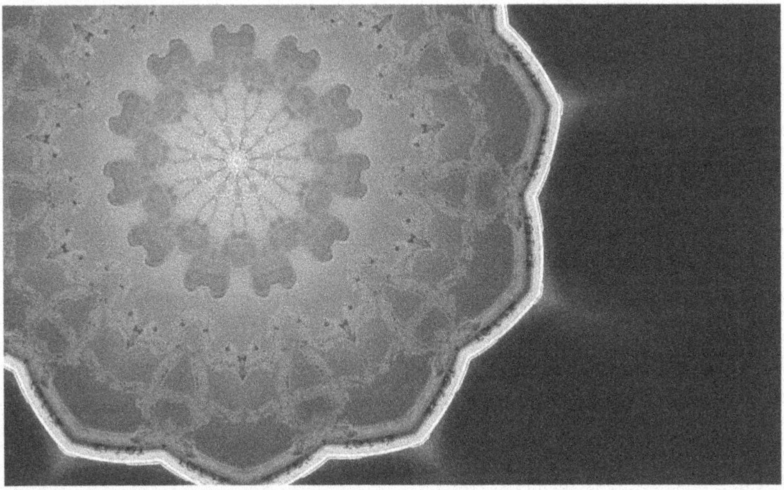

Figure 5.5.6 Lilwatan (To the Homeland), video still image, 2017 – Jude Abu Zaineh juxtaposes Middle Eastern motifs with western contemporary techniques to produce kaleidoscopic renderings of cultured bacteria growth, animated on old television sets

Source: Jude Abu Zaineh

The Fair was presented in the Science Centre's Great Hall as part of the Centre's Inventorium initiative. Staff worked alongside artists and volunteers to facilitate the experience of 14,000 visitors, who could also purchase artists' work and kits to extend their creative explorations at home. Projects presented at the Fair were extended through an online exhibition: www.sciartcenter.org/tech-art-fair.html.

A total of 130 artists and volunteers participated in the conception and presentation of the Fair. Artist submissions – solicited through the Science Centre and partner-driven promotion over a six-week period – were collected on the arts and culture platform www.huutaart.com. Projects were then selected by a jury drawn from the partner organisations (Toronto's Propeller Centre for the Visual Arts, Daily Goods Design Labs and New York City-based SciArt center, with additional support from InterAccess and Art the Science) and led by the centre. Installation was supported by dozens of centre staff: from research, design and fabrication, to marketing, programming and floor hosts, all in collaboration with artists and their volunteer assistants.

Outcomes

The fundamental goal of the TechArt Fair was to experiment: could the centre partner with arts and tech-focused communities to co-create an event in which artists could expand their networks, sell their work, and engage visitors with art and science?

The experiment produced three powerful results:

- It connected art and creativity with science, tech and innovation – for both visitors and participating artists;
- It expanded the centre's network through an open-call and juried process;
- It impacted the centre's internal approaches to experience creation and delivery by giving direct access to external creators. One striking example was the artwork *Trappist Sounds* by Matt Russo, which enticed visitors to jam together using musical tones created by data about the orbits of the TRAPPIST-1 planetary system. The observed success of this experience with visitors led to an ongoing collaboration with the artist to identify opportunities to expand his work, including potential collaboration with the centre in Toronto's Nuit Blanche festival.

Selection criteria for works of art entering science communication institutions are key to the success of initiatives. Art projects that 'reveal creative process, inspire awe and excitement and spark interaction, collaboration and co-creation with visitors' are more successful in changing participant's perspective on a long-term basis.

5.5.7 Encouraging long-term or top management collaborations

The EU funded Traces Project (2016–2019) analysed ways in which artistic and ethnographic research can jointly approach 'contentious heritage' and convey it to the public. Findings include recommendations on how collaborative work among different fields of expertise can be effectively conducted. Another EU funded project – The European Digital Art and Science Network – gives practical tips on artist residencies in science research institutions. In both cases attention is brought to the necessity of long-term collaborations, that can affect institutions more deeply.

Other attempts to produce deeper changes have worked less on the time factor and more on the responsibility-level of the actors involved. These experiments foresaw the inclusion of artists at the level of directorship, which entailed some issues connected to a top-down approach, but also prompted the involvement of different museum departments – something that projects involving smaller groups of professionals in an institution (such as educators or curators) may not achieve unless they are given much more time to penetrate from one sector to the other.

Case study: Pavilhão do Conhecimento – Ciência Viva, Lisbon (Portugal)

Gonçalo Praça, Project Manager

Ciência Viva is an agency for the public awareness of Science and Technology in Portugal, whose associate bodies include public institutions and research laboratories. Ciência Viva is responsible for the creation of a national network of 20 interactive science centres in Portugal.

Today I'm the Boss

An experiment in shared curatorship that put the programming of Ciência Viva's main science centre in the hands of an artist for one day. This experience inspired regular collaborations with artists, engaging some of them in the "public board" of the institution and initiating the practice of including artistic components in the development of most recent exhibitions.

Description

In 2010, Ciência Viva invited professionals from music, design, fashion, circus, dance, theatre and literature, the director of the Finnish science centre Eureka, and the team of the Pavilhão do Conhecimento explainers to each head the science centre for one day. These invited "directors" were responsible for the programme of the last Saturday of each

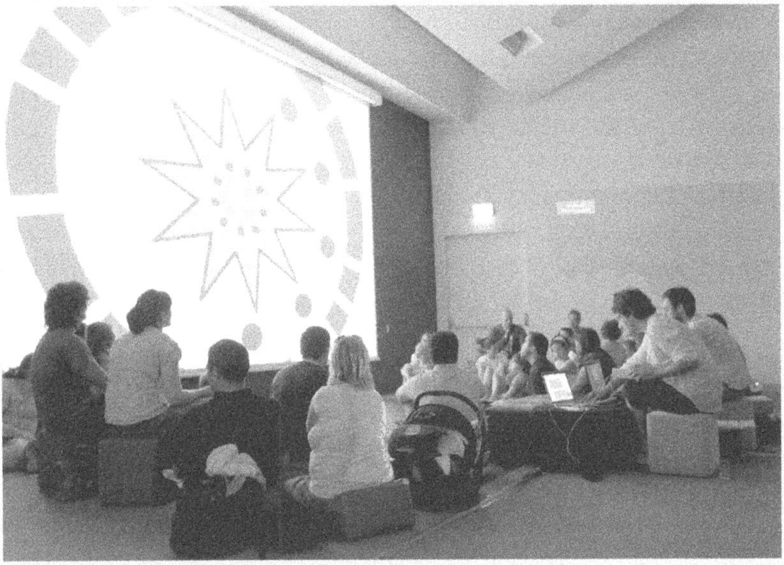

Figure 5.5.7 Sound with colours: electronic music for babies live performed at Pavilhão do Conhecimento – Centro Ciência Viva, 31 July 2010. The event was part of the Universal Day of the microorganisms, curated by Mark Deputter and Patricia Portela, and focused on the "science out of the reach of the naked eye", Ciência Viva

Source: Photo by Rui M. Leal

month. The directors designed the programme for their day, prepared budgets for the activities and events, selected and invited collaborators, and hosted meetings with a team of Pavilhão to delegate tasks for the implementation of their agenda. Each director also wrote an "editorial", explaining his/her vision for the day. During these dates the venue was open well over normal working hours for €1 (a substantially reduced fare) and activities took place all over the science centre, including exhibition galleries and outdoor spaces as well as areas that would normally be closed to the public.

The activities offered to the public were substantially different from those taking place on an average opening day and included storytelling sessions, workshops (sewing, plant growing, creative electronics, etc.), live music, dance and theatre performances, artistic installations, talks and debates (including web conferencing with science centres abroad and the production of the book "Design Meets Science").

Regular staff members were involved almost only in logistic tasks and as scientific reviews of the programmes.

Invited directors were Mário Laginha (pianist) and Cesário Costa (director of the Lisbon Metropolitan Orchestra); Bárbara Coutinho (director of the Lisbon Design Museum) and Henrique Cayatte (graphic designer); Francisco Rebelo (musician) and Carlos Martins (sax player and artistic director of Sons da Lusofonia Orchestra); Teresa Ricou (head of Chapitô school of circus and performing arts); Mark Deputter (director of the municipal theatre Maria Matos) and Patrícia Portela (theatre director and writer); Per Edvin Persson (director of the Eureka Science Center, Finland); Patrícia Reis (writer and journalist); and Rui Horta (choreographer).

Outcomes

Ciência Viva considered art to be attractive for audiences that were not familiar with the science centre's activities; well-known, influential artists were invited with this idea in mind. Ciência Viva also wanted to experiment with new curatorial skills and experiences that could be adapted and developed in the long term.

Audience numbers were not significantly greater than those of normal weekends yet the attraction of different types of visitors was commonly recognised as an important achievement.

Anecdotal evidence suggests that the session organised by Pavilion of Knowledge explainers attracted more visitors and engaged them for a longer period of time. Team members believe this was due to their deeper knowledge of the public, being more attuned to their interests, more informal and less "erudite" (compared to the invited artists).

Frictions between the different cultures of science and the arts were evident. For instance, the comments made by Pavilhão staff on the script of a surrealistic performance enacting a "micro-encyclopaedia of micro-organisms" noted: 'We don't quite get the scientific "truth" on which this is based'; 'That designation is no longer used [in science]'. "Science" was often perceived by staff as a layer juxtaposed on top of the artistic and cultural contents prepared by the directors and their invitees.

More long-term interaction between the invited "directors" and Ciência Viva staff could have allowed for a more intimate relationship between science and arts in the frame of communication and education. In fact, this prolonged relationship has proven to be an essential component in long-term outcomes such as the invitation for a number of "directors" to become members of a new "public board" of Pavilhão do Conhecimento – a participative body which gathers professionals from different fields, students and local authorities to contribute their point of view on Pavilhão's programme and strategy. It is fair to say that the "public board" embodies the spirit of the *Today I'm the Boss* experiment, with deep repercussions in the governance of the science centre.

5.5.8 Learning by making

Many science centres and museums have crossed paths with the Maker Movement in different ways: they may have spoken about it in exhibitions on innovation, or invited members of the movement in talks and programs, or created situations for start-ups and professionals to meet with their audiences, etc.

Many have also decided to set up Maker Spaces within their own walls to convey the empowering message of taking into one's hands the prototyping and production process instead of relying on others.

In any case, maker spaces and science centres share the idea that there is a profound educational value in "learning by doing". Experiential Learning – which is traditionally traced back to the work of John Dewey and later on to that of David Kolb – highlights the importance of presenting learners with real-life problems and promotes learning processes in which knowledge is created through the transformation of experience itself.

Case study: STEAMLabs – Maker Place and Maker Bean Cafe, Toronto (Canada)

Andy Forest, Co-founder

STEAMLabs is a young community makerspace where people of all ages and abilities come together to practice digital fabrication, coding, hands-on making and electronics. It focuses on not for profit programs emphasising creativity, critical thinking, collaboration and communication.

Maker Spaces

Two spaces in the Ontario Science Centre, a Maker Place and a Maker Bean Cafe, that encourage visitors to play, create and connect with technology and each other. Technology education, in the form of making activities, encompasses aspects from the fields of design, engineering, aesthetics, art and science. Empowering participants to be protagonists of the creative process means offering them challenges and learning opportunities that strongly relate to their personality and interests.

Description

Having run a kids technology makerspace in Toronto for eight years, STEAMLabs was invited to co-create the "Maker Place" at the Ontario Science Centre. It is a space for both kids and adults who wish to get hands-on with digital fabrication technology, coding, making and electronics. Visitors design things on paper to be laser cut on the spot. Robots, critters, vehicles, jewellery and other creations are assembled, decorated and tested in challenge areas such as a garden with fans to blow insect critters to the flowers. By optimising the workflow, using inexpensive materials such as wood and cardboard, and self-explanatory support materials two to three staff can serve up to 40 makers at a time, involving hundreds of thousands of visitors since the opening in 2017.

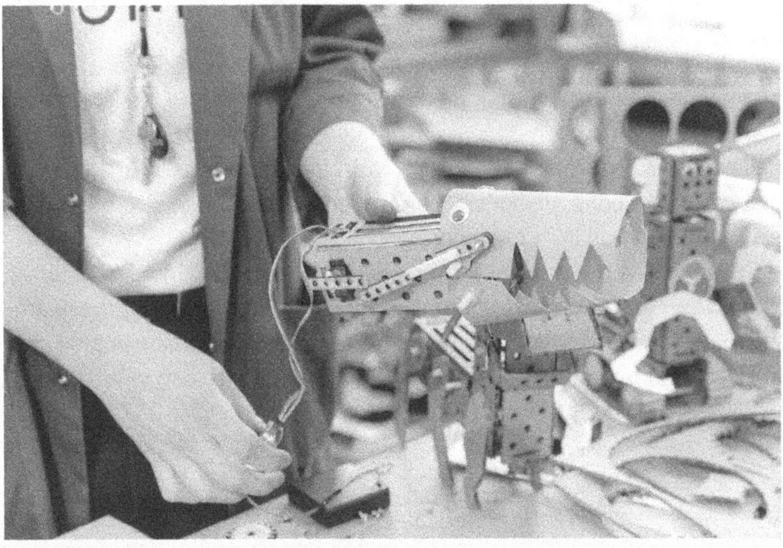

Figure 5.5.8 Visitor creating crocodile and other cardboard robots at steamlabs Ontario Science Centre makerspace

Source: Ontario Science Centre

To complement this free experience for visitors, the adjacent "Maker Bean Cafe" offers a more guided, paid experience. After initial brainstorming with staff about fun things that can be made, objects are designed on a computer, introducing visitors to digital design. Projects are then laser-cut on the spot as a take-home souvenir of the experience. The activity generally lasts 15–30 minutes and takes place in a cafe setting, where visitors can enjoy hot drinks while being creative. There is a small fee per item created that serves as a unique souvenir of a visitor's time at the Science Centre.

Outcomes

The experience brought together leading organisations in technology education and created a "third thing" that none would have been able to do on their own.

The Maker Bean cafe, in addition to providing a more hands-on experience to complement the self-serve Maker Place, also offered a sit-down place for people to relax and share their experiences. Families split up and went to different areas on their own according to their interests, and reconvened at the cafe to engage in making (more than 90% of the visitors actually created something), sometimes lingering for hours.

For members and regular visitors to the Ontario Science Centre, these spaces have provided a place to do something new on every visit. Many members return week after week to make new things. Repeat visitors also stepped out of the bounds of planned activities and designed their own. People brought in designs they made at home to be laser cut and have mashed up materials and tools from other areas to make more elaborate creations. Visitors have used these spaces to express themselves, to express their identities.

When kids are empowered to be makers, amazing things happen in their lives. By being creators instead of just consumers, they are learning to be the innovators of the future, today. Their confidence in their own abilities allows them to make the change they want to see in the world.

6 Art, inclusion, controversy and imagination

From facts to values

Science engagement institutions that aim to present contemporary scientific and technological research must necessarily face questions about future social/ethical implications and scenarios. Much contemporary science research can be recognised as controversial and the uncertainty of its outputs and applications opens up debates that may benefit from imagination as well as from a good dose of self-criticism.

New kinds of institutions as well as new curatorial viewpoints are now basing their programs on sci-art collaborations. Very often these venues and professionals choose not to present solely scientific or technological facts, but rather to connect them with themes that have broader social relevance. Their main objective is not to explain, but to stimulate interpretation, reflection and productive doubt.

In their slightly anarchic interview, Mike Stubbs and Mark Wright from FACT in Liverpool (UK) invite museum practitioners to find new ways to engage the audience in activities that nourish their creativity and personal representation and interpretation of a topic. In order to do so, they advocate for a less rigid relationship with the idea of truth, and highlight the importance of the concept of "uncertainty" in both artistic and scientific research. This attitude can ultimately reinforce an image of science as a cultural product. From the museological point of view, a more rich ecosystem of inter-connected disciplines can 'enable the public to engage, interpret and creatively respond'.

From an artistic perspective, artists Oron Catts and Ionat Zurr – whose research at SymbioticA in Crawley (Australia) is focused on life forms that overgrow the common concepts of normality and nature – present a reflection on the role played by context and its relationship with the artworks and its viewers. The authors discuss the frames in which artistic living artefacts are displayed, and investigate the relationship between museums and artistic research.

Indeed, museographic choices reflect also the spirit of the time. Cultural institutions reveal in each action their social, political and ethical perspectives. The way objects and topics are selected, displayed and narrated are all fruit of conscious or unconscious decisions. Art may strengthen, smoothen, contradict, re-think, judge or interpret these processes. In her chapter, Bergit Arends describes the experience of collaborating with artists from diverse cultures that have re-interpreted some of the collections at the Natural History Museum in London (UK). Arends describes how the artists' new perspectives provided an opportunity to reflect and sometimes challenged past and present curatorial practices.

Case studies in this section have been selected because of the controversial themes they have chosen to engage with, the different ways they have found to question the image of science, or the curatorial attitudes that they have aimed to highlight and unveil.

6.1 Criticality, imagination and interaction

A new basis for art science curation

Mike Stubbs and Mark Wright

Introduction

Contemporary curation and engagement operate within an increasingly divided society. The cultural condition is one of social, political and intellectual fracture. Art and Science have an important part to play in the creation of new forms of discourse within diverse, knowledgeable and empowered audiences, yet enabling space for ambiguity and the imagination.

How can we commission and present art and science in a way which raises awareness and challenges our cognition of the world? How can we move beyond discredited grand narratives, which embed misconceptions of science and art as always being separate and the domain of a small elite? How can we pursue an agenda of hybrid practice exemplified at the heart of new media, technology and science? How can we instil an understanding of art, technology and culture as forms of embodied knowledge and individual literacy?

This chapter explores an approach to art-science collaboration, curation and public engagement, which has developed at FACT the Foundation for Art and Creative Technology in Liverpool (UK). Here, a rich practice has emerged which raises issues concerning the place of science in wider culture, public understanding of the role and nature of science and how fruitful collaborations between art and science can be created which provide provocations around these issues.

In this chapter Mike Stubbs (CEO and Artistic Director of FACT until 2018) and Mark Wright (FACTlab Research Director) engage in a joint attempt to answer some of the complex questions raised by the very idea of presenting art-science collaborations in a space of public engagement.

Conversation

Science is not just about rote knowledge. It is about a process of reflection, action, error and discovery and this process is familiar to artists. To be unaware of this process is to be unable to truly understand science or art, which is a state the public visiting art or science engagement institutions are often trapped in. Even though Art and Science seem on the surface to be about the visual and spectacle, they are both concerned with a deep curiosity of the unseen which may be conceptual, abstract or indirectly observable.

Richard Feynman said that, most of the time, he experienced a state of not knowing and playing and being wrong. What role does imagination play in scientific research, in art practices, and in the way the two are generally perceived?

Mike Stubbs (MS)

The sense of being relaxed with not knowing is important. If science is shrouded in mystery and specialist language, which it has to be to a certain extent, how can we penetrate its physical spaces and feel confident enough to explore them for ourselves? Something similar happens to scientists with art. If you visit a painter's studio you realise they have painted all day every day for the last 40 years. How does that compare to a scientist at CERN spending 40 years trying to find the Higgs boson? They are not dissimilar. Their motivation is based on a sense of curiosity and an ability to imagine something that is not already here. So quite a bit of what is written in science fiction is not about the future but about what is actually happening.

With predictive technologies things get closer and closer to us as we live with the illusion that we know everything, or that we have the ability to understand the concrete world in real time, whether that is through IoT (Internet of Things) technologies, metadata combined with high speed networks, as a process of prosthetic augmentation of the self. Like painting or good old-fashioned scientific research, some of these systems are being replaced by forms of automation, this gives us the idea that we are more distant from the thinking process.

Mark Wright (MW)

Uncertainty is what gives all knowledge its humanity. To be disconnected from this uncertainty, as the public often are with science, leaves us unable to place that knowledge appropriately within our own personal cultural context.

Without a feeling for uncertainty, knowledge seems cold, threatening, meaningless or infallible. This leads to a sense of dissociative confusion or dogmatic acceptance or rejection. We need to find new ways of connecting people to an appropriate sense of uncertainty, doubt and possibility through which meaningful encounters are made possible. This sense of uncertainty is a prerequisite for the public to feel empowered to apply their agency and imagination.

The creation and sharing of knowledge unfold in a state of uncertainty and flux. During this creative process artists, as well as scientists, enter an embodied participative status of becoming. This experience can be very exhilarating and rewarding. Knowledge created by this process is like a fossil, a trace left behind by embodied creative interaction. Again, the public are unfamiliar with the uncertainty associated with this process and this state of being. Instead they see the fossil and are unable to bring it back to life. Similarly, sheet music is not the same as music as it is played and heard.

Knowledge is not an artefact, knowledge is performative. One thing music, art and science all have in common is they are 'played'. They are performed and explored through an embodied process of playful improvisation.

We therefore need to find new ways to enable the public to experience this creative participative status, both in the creation of new ideas and their imaginative reinterpretation, extension and re-contextualisation.

Science is usually thought of as being either "correct" or not. Instead, what if we were to approach it in terms of different cultural interpretations from an ensemble of voices?

MS

It is interesting to see how people who study science histories can make more speculative connections between science, religion, belief and other indigenous beliefs of how the world was formed and how we experience it. In our experience, the artists that have worked in combined residencies at FACT and CERN have been respectfully irreverent, deeply questioning the *raison d'etre* of these institutions.

For example, the two-artist collaborative "Semiconductor" is clearly interested in the epistemologies of science processes and in the people that actually do science. Yunchul Kim on the other side is more interested in an intimate understanding of the research material itself. Yet in both cases their work enables audiences to experience places, such as labs and research facilities, that otherwise might feel detached and institutionally predetermined. There is a general need to gain access to these spaces through other people's experience, whether through fiction, documentation of a process or the recounting of a tale.

MW

The principal reason science tends to dominate cultural discourse in this way is that, in various ways and without realising how, we get trapped inside its ontology.

An ontology consists of a set of concepts and categories and their relationships which can be used to define a subject or cultural domain. An ontology of science, for example, may consist of many things such as; "physical phenomena", "abstraction", "theory", "experiment", "observation", "data", "analysis", "refutation", "paradigm" and so forth.

To understand and practice science as scientists do is, in a sense, to learn, operate and expand the ontology from within. To educate people in order to become scientists it is also appropriate to adopt this "internal" perspective. However, confusion can arise if we apply this "internal" approach to public engagement or as an all-encompassing world view.

The principal aim of scientific public engagement should not just be to teach the form and use of the ontology of science, but to also enable the public to relate that ontology to wider culture.

It is unsurprising, given the obvious transformational power of science, that some readily apply its ontology beyond the practice of science itself. Conversely, it is also unsurprising that other secular and religious aspects of culture lay claim to parts of science and such appropriation and contestation is a normal part of cultural discourse. However, it is important to see that the ontology of science and its expressive power have limits. In particular, subjective experience is outside its domain. Science can tell us what a human is but not what it is to be human. Science can tell us how we behave or inform our behaviour but not how we should behave with respect to each other and the world.

Science should be seen as one important aspect of culture, one ontology amongst a network of others to be related and explored.

In such a landscape, different cultural interpretations and voices have space to open up. We therefore need to go beyond the concept of trans-disciplinarity to trans-ontological methods, which recognise difference, plurality and relations as fundamental.

Academic and formal settings are beginning to investigate whether textual descriptions are "enough". How does contemporary socially engaged art give people the means to experience and construct content for themselves?

MS

Clearly there are increasingly diverse mechanisms that enable people to have more first-hand experience. If we think of it, literary traditions went from the third to the first person; entertainment went from theatre, to cinema to TV; computer games moved on to Virtual Reality experiences, creating hybrids of shared and private spaces. In our recent work with BeAnother-Lab from Barcelona, there is the use of Virtual Reality and Augmented Reality to give the direct experience or illusion of being inside someone else's body. It is another form of representation just like projected media but feels a lot closer to a representation of being that other person.

Our current condition, on the verge of the neural networking, seems to promise deeper shared experiences of consciousness. Yet this is all but an end point, which brings us back to the importance of science fiction and imagination. As we develop new technologies that provide new subjective and inter-subjective experiences, we must constantly find ways to ask ourselves: what then?

MW

Academics have increasingly become involved in new forms of enquiry that involve creative interaction with real world social-technological systems. Methods such as action research and the ubiquitous rise of co-design are examples of this.

A principal reason for this is some forms of knowledge are situated and embedded in these systems physically or as tacit knowledge and social relations rather than being readily at hand for observation. It is only through interaction that they are revealed. Also, if we follow an approach of creative, critical speculation, we are creating entirely new embodied and situated systems in which new meaning, relations and consequences unfold in an unpredictable manner as we creatively form these systems in vivo. The study of culture in such a way requires a new trans-disciplinary approach. Hybrid methodologies involving concepts from design, sociology, ethnography and arts practice have emerged to support this process. As a consequence, the concept of publishing has also been revisited. Conventional analysis and case studies are augmented with information concerning ways to recreate immersive experiences and community interactions.

Conventional approaches to scholarship are well suited to their role of exploring, deepening and interpreting ontologically distinct silos of traditional subject boundaries. They are not suited to 'trans-ontological' research. Science alone sees such enquiry not as science but the "application of science", arts and humanities alone see such enquiry not as the study of culture but an intervention into culture.

We do not call for the abolition of traditional 'silo' scholarship as this provides the deep knowledge in science and the arts and humanities, which is the source material for our approach. What we call for is to augment the traditional with trans-disciplinary methods. This requires a new form of practitioner who is able to peer more deeply into these silos and contextualise their meaning into wider culture.

Socially engaged arts practice has been influential in pioneering such methods, as they are spaces in which the social meets the digital and scientific aspects of culture. Practitioners in this field have developed methods similar to that of academics. However, pedagogical, social and political principles apply which widen the aims of interaction to learning, agency and empowerment.

Socially engaged art practice dealing with "speculative futures" and "social dreaming" have produced works that aim to create new forms of public engagement. What kind of cultural spaces can arise from this new idea?

MS

I think we are only just learning what that might be and to use new language in new settings. There are lots of interesting developments in terms of science futures and speculation. What I believe is that there is a need for collaboration between science research institutions and audience facing places that engender a sense of openness and sharing, that leave space for ambiguity, magic and realism.

MW

The essential element that art provides for engagement is not illustration or explanation but criticality. At FACT we think of public engagement in terms of the concept of "Literacy". To live in a society where science is so important citizens must be scientifically "Literate". Literacy usually means to be able to read or write. In this wider cultural sense "Literacy" means to both understand and act. We use art, not for mere illustration or dissemination of science, but as a critical provocation to enable the public to engage, interpret and creatively respond. We put these two elements of criticality and literacy together to create new forms of engagement.

People must grasp the processes and uncertainty of science. Being presented with dogmas builds kickback and resistance to statements. This goes back to the importance of being in connection with a state of doubt. Also the focus on objectivity sheds everything else, the political, the social all become invisible because of this focus on the physical world and only that which is objectively knowable rather than the subjectively interpretive. Having spaces that put aspects of the material world recounted by science back together with interpretations helps to put things into perspective and see that science is a part of culture that acts through frameworks of power. The public create their own understanding, meaning and voice through experiences which are embodied, speculative and imaginative.

A note on the role of embodiment in art and science

The concept of embodiment might be interpreted as a unifying perspective to draw our reflections together. Embodiment is a philosophical position, which takes the body and the physical world to be as much a part of our cognitive apparatus as the brain (Clark, 2011). When the first ape picked up a twig to poke a termites' nest, it ceased to be merely an animal and became a cyborg, part animal and part technology. To be a cyborg is not just to have a tool to achieve a new task but to have an extended cognitive landscape. Human evolution can be seen in terms of a multitude of new embodiments, which transform our cognitive, social and cultural awareness.

Language and mathematics, its formal derivative, are also technologies derived from embodied interaction (Lakoff & Núñez, 2001).

All arts media are technologies. Oil paint is a technology, which allowed artists to work more flexibly. In this perspective we can include new media as a natural progression of this cultural evolution without falling into the trap of fetishising the technological. Scientific instruments are technological embodiments, the microscope, telescope offer new embodiments which allow us to see things which could not be previously seen. More deeply, when we consider radio telescopes and particle colliders, it is not just that we see something hidden but that embodied interaction is what brings perception into being. Language and mathematics provide a further form of conceptual "seeing". From this perspective the cultural significance of technologies, mathematics and scientific instruments, is that they create new embodiments, which bring about new forms of knowing and being. We are now entering a phase where the concept of embodiment extends beyond the augmented individual to internet scale computation, post human and non-human ecologies of interaction.

New technologies such as computer simulations, projectors and virtual reality, bring a new class of embodied experience into existence. The abstract can be made visceral in new experiential realities. Visualisation taps into the old visual pathways of the brain to bring insight to data from images, not just through aesthetics but enhanced forms of cognition (Feraco & Erdt, 2018)

This relationship between digital media, culture and remediation has profound consequences for curation and the concept of the archive and database as noted by Hayles (2012). The archive is transformed from a passive repository to an enactive space for experiential narrative creation and sharing.

Furthermore, the concept of embodied cognition underpins the theory of enactive learning and constructivism. Learning and cognitive development are not just processes of logical deduction but of action in the world.

The very need for the new approaches to curation and engagement advocated here using methodologies of critical provocation, imagination and co-design, are predicated on the view that all human relations are embedded and situated physically and socially and that meaning unfolds through embodied interaction. It is also the source of the embodied participative status of becoming and its uncertain potential which artists and scientists enter in the knowledge creation process and which we seek to engender in the public through critical, imaginative creativity (Dourish, 2001).

As contemporary curation looks to highlight process, networks, algorithms, global computation, non-human and post-human (Graham & Cook, 2010), our concept of embodiment moves from a single person, a single body using a single tool to networks and ecologies of embodiments physical, and social.

We believe it is possible to use this framework of embodiment combined with the criticality of arts practice to encourage a deeper connection between art and science rather than just to select art that illustrates or presents a technological spectacle to engender a critical imaginative unfolding of understanding through enactive learning.

These new contexts exist where the materials, processes and philosophies that shape a new approach to understanding the world, are formulated in a space of imagination, the fictive and belief along with proven data and beyond traditional science engagement.

Shaping new curatorial and engagement practices

FACT has developed long term engagement projects which build trust while maintaining artistic integrity. There has to be respect for the participants and the process from all sides. FACT conducts engagement in many forms but have a particular interest and expertise in long term deep engagement. Participants of such engagement are taking a risk of trusting the artists that they will not feel inadequate or foolish. Many communities are vulnerable such as those with social, physical or mental health challenges and their welfare has to be ensured. Deep trust can only be built up over a long time. It is important that the practitioners show respect and a clear commitment to the community. The practitioners must also be prepared to take risks and be prepared to change their approach as things develop.

Artistic integrity is also an important aspect of engagement. So-called socially engaged art has been critiqued for choosing between aesthetics and social activism while falling short on both (Bishop, 2012). A central strategic goal of FACT is international excellence. Our exhibitions on art and science and the CERN Collide programme attract internationally recognised artists as well as emerging and mid-career artists of the highest order. This commitment to professional excellence in both the art and science is important.

In socially engaged practice there is, by definition, a collaborative public element. One way that artistic integrity can be maintained in this sense is by careful framing of the aesthetic context by the artist. Other artists may encourage more agency by the community, but it is still possible to create exceptional work. Another approach is for other artists and engagement professionals to aid the public to respond to a primary work in the gallery either directly or conceptually.

We have shown there is a history of socially engaged practice involving many types of arts institution. We believe that art and science curators can learn from this history and cross-pollinate new ideas of participation.

Theories of pedagogy are relevant to social engagement both in terms of political awareness, where education is equated by practices of freedom rather than the banking of facts (Freire, 1970) and constructivist approaches which foreground understanding through enactive doing rather than passive presentation based on the developmental psychology of Piaget (Piaget, 1954), its connection to cultural and dialogic processes (Bruner, 1990) and explored in socio-technical educational contexts by Papert (1980).

Publication is no longer confined to an exhibition catalogue, analysis or case study. It now extends to the sharing of resources and recipes for the recreation of communities of experience. This has been enabled by the World Wide Web (McGann, 2001) and the rise of the digital commons and commoning, the creation of free shared or open source resources, extending beyond software to hardware and other physical and social resources (Baldauf & Gruber, 2017).

Socially engaged artists have recognised the need to, not just document, but share the processes, methods and actual artefacts which embody complex frameworks for human encounter and discovery (Re-Dock, 2017).

Maker culture is increasingly visible in art and science settings and is also implicated in encouragement of children into STEM subjects, which is an important context for art and science collaborations and exhibitions. Making certainly seems a natural partner to socially engaged art, constructivist learning and speculative design.

FACT has integrated making into all exhibitions and engagement using FACTLab which is a production, engagement and research space embedded within FACT itself.

It is interesting to note that recent research (Blikstein & Worsely, 2016), warns against light demo style workshops, the separation of abstraction and practice and the operationalisation of maker learning as a STEM gateway. Instead it is important to support learners through to a level of fun but tough projects which combine abstract concepts, practical outcomes and empowered creativity as their driving force. FACT is designing exhibitions and engagements with our publics, which embody these principles.

Conclusions

We have argued that science and art are intrinsically uncertain processes that operate in a state of doubt. Something which the public is not often aware of and with which we must reconnect. Another point was made that science should be seen holistically as a cultural phenomenon and not just in terms of knowledge production. These two issues come together and call for a new curatorial approach which sees audiences as free cultural agents. The public is not just a receptacle for what science states to be true: in a society where science plays an increasingly important role, it is composed of active citizens that fully engage in a dialogue through which the future unfolds.

All arts and science institutions are aware of the importance of public engagement. In the arts this refers to the inclusion of diverse audiences beyond a privileged elite. In art institutions, engagement with the public has extended beyond public exposure and understanding of art to part of the artistic process itself. This was noticed by Bourriaud as an aesthetic development (Bourriaud, 2002) and documented by others in its many diverse forms (Thompson, 2013).

Curation of art and science has long moved on from the display of artefacts in the context of a single narrative to one that mirrors some of the constructivist viewpoint. However, there may be more to learn about meta-curation of experiences which precipitate enactive dialogic experiences.

Recently, concepts from design have entwined with those from arts practice so that expression, creativity and learning combine in speculative imagining of the future both on a social and individual level (Dunne & Raby, 2014). Engagement is far more interesting using this model than one purely about technology or skills acquisition.

This leads to citizens who are actively empowered participants in their own cultural world.

Bibliography

Baldauf, A. & Gruber, S. (2017) *Spaces of Commoning – Artistic Research and the Utopia of the Everyday*, Berlin: Sternberg Press.

Bishop, C. (2012) *Artificial Hells: Participatory Art and the Politics of Spectatorship*, London and New York: Verso.

Blikstein, P. & Worsely, M. (2016) Children Are Not Hackers: Building a Culture of Powerful Ideas, Deep Learning and Equity in the Maker Movement, in: Peppler, K., Halverson, E. & Kafai, J.B. (Eds.) *Makeology: Makerspaces as Learning Environments*, 1, 64–79, London: Routledge.

Bourriaud, N. (2002) *Relational Aesthetics*, Dijon: Les presses du reel.

Bruner, J. (1990) *Acts of Meaning*, Cambridge, MA: Harvard Press.

Clark, A. (2011) *Supersizing the Mind: Embodiment, Action, and Cognitive Extension*, Cambridge, MA: Oxford University Press.

Dourish, P. (2001) *Where the Action Is: The Foundations of Embodied Interaction*, Cambridge, MA: MIT Press.

Dunne, A. & Raby, F. (2014) *Speculative Everything: Design, Fiction and Social Dreaming*, Cambridge, MA: MIT Press.

Feraco, A. & Erdt, M. (2018) How Visualisation and Interaction Can Optimize the Cognitive Processes Towards Big Data, in: Mehdi Khosrow-Pour, D.B.A. (Ed.) *Encyclopedia of Information Science and Technology*, 4th ed., New York: Routledge.

Freire, P. (1970) *Pedagogies of the Oppressed*, New York: Herder and Herder.

Graham, B. & Cook, S. (2010) *Rethinking Curating: Art After New Media*, Cambridge, MA: MIT Press.

Hayles, K. (2012) *How We Think: Digital Media and Contemporary Technogenesis*, Chicago, IL: University of Chicago Press.

Lakoff, G. & Núñez, R.E. (2001) *Where Mathematics Comes from*, New York: Basic Books.

McGann, J. (2001) *Radiant Textuality: Literature After the World Wide Web*, Basingstoke: Palgrave.

Papert, S. (1980) *Mindstorms*, New York: Basic Books.

Piaget, J. (1954) *The Construction of Reality in the Child*, New York: Routledge and Kegan Paul.

Re-Dock. (2017) *Critical Kits and How We Use Them*, Itinerant: Torque Editions.

Thompson, N. (2013) *Living as Form: Socially Engaged Art from 1991–2011*, Cambridge, MA: MIT Press.

6.2 Shifting meaning, shifting contracts – biological arts and evolving museum ethics

Oron Catts and Ionat Zurr

Writing from the point of view of the artist, we will examine the attitudes, implications and procedural aspects of presenting critical and reflexive biological art in different cultural and educational contexts. The authors will draw from their 22 years of experience in developing and presenting artworks which utilise manipulated biological systems contextualised to examine complex relationships with the concept of life itself. Presenting these works in a range of cultural and educational contexts such as art galleries, design museums, natural history museums, science centres and other institutions, the authors will reflect on the shifts in meaning and concerns that each context generates.

Following the contestable ethos of the museum as a "Safe Place for Unsafe Ideas" (Cameron, 2008) the question we ask is what happens to critical art when it is contextualised and/or situated in a "safe place" and what are the strategies that can be employed to make these "unsafe" ideas effective and to whom? Drawing on some of Tony Bennet's (1995) ideas (who viewed the modern museum as a place of discipline for society's evolving morals) we look at ways of conduct and reforming behaviours and ask: what is the role of the critical artist? Who does the museum's context discipline beyond the living and semi-living artwork – is it the artist? The audience? Or the museum itself? Biological art adds further complexity to existing principles of all modern museums, be they cultural, natural history or art museums, particularly in relation to the conservation and preservation of objects in perpetuity.

How biological arts are read across different museum types

Here we would like to tease out some of the multiplicity present in showing artistic and techno-scientific Bio-Objects (Vermeulen *et al.*, 2016) shown in different museum contexts. Specifically, we would like to "put on display" two phenomena, which appear simultaneously:

1 There is an interesting shift in the relationship between, and fusion of, natural histories and cultural history narratives. As a result of the shift

in biology from a scientific to an engineering pursuit, life and living systems are becoming raw materials for human ends, and as such cross over form natural context to technological and cultural objects.

2 The museum becomes implicated in the fetishisation of technological approaches to life, to the end that it may overshadow the context in which life operates.[1]

These inter-related phenomena will be illustrated through discussions of specific bio-artefacts; both with scientific and/or artistic motivations.

Neolife: natural specimens or technological items?

In a recent visit to the National Museum of Scotland in Edinburgh, we encountered an uncanny display. In some ways it followed aesthetic conventions common in natural history museums, but some details as well as the context of the display were drastically different. A stuffed (taxidermied) animal was put on a revolving circular stage in a somewhat heroic position. It was a white sheep. The sheep was looking straight at the viewer; standing in scattered hay on the floor and plastic replicas of what looked like faeces.

The stuffed animal did not represent its species, but was rather a unique, historical, techno-scientific wonder. It was not placed in the natural history wing of the museum, rather it stood near steam engines and computers in the science and technology (or "innovation") section. Dolly the sheep, the mammal cloned from an adult somatic cell, was put on a circular rotating "petri dish" surrounded by other technological items.

An interactive digital display nearby asks: "Why is Dolly important?" and answers:

> Dolly was important because she captured the public imagination. The idea that there might be an exact copy of oneself somewhere in the world is a theme that has been often pursued in science fiction and the prospect that it might be possible to clone a human being has excited a lot of speculation and interest. Likewise, plans to clone extinct species such as mammoths have attracted a lot of publicity, but at present such ideas must remain, like *Jurassic Park*, firmly in the realm of fiction.
>
> (National Museums Scotland, n.d.)

How does a museum display Dolly's techno-scientific hype and wonder if she looks just like any other sheep? Is it by merely moving her from the context of the living world to the realm of human-constructed machines? Is it by fetishising the technological approach to life rather than emphasising Dolly as a sheep with her own agency and *umwelt*?[2] This is a prime example of what we call *Neolife*.

The Wellcome Trust collection have gone one step further in terms of reducing Dolly from a whole sentient animal to a fetishistic icon by displaying not a vial of Dolly's DNA, but rather her "real" faeces, preserved and on show (Rose & Rose, 2007).

While Dolly's exhibit follows the contestable ethos of the museum as a "Safe Place for Unsafe Ideas" by positioning a "normal" looking sheep among machines, there are isolated, anecdotal examples of more contestable displays. One of these can be found at the *P. Stradins Museum for History of Medicine* at Latvia Riga. This is a taxidermy of a "two-headed dog"; more accurately – a dog's head transplanted onto the neck of another dog. The display is a relic from the experiments in transplantation of the Russian scientist Vladimir Petrovich Demikhov. Demikhov started these experiments in 1954 as part of a continuing series of experiments in surgical techniques with the ultimate goal of learning how to perform a human heart and lung transplant (which was successfully performed by Dr. Christiaan Barnard in 1967[3]). The artefact in the museum has a label attached to the display that reads: 'The exhibit at the museum was produced in 1966 especially for the Pauls Stradins Museum of the History of Medicine. The dogs lived for four days after the operation'. The label may suggest that the procedure of transplanting a puppy head on another dog was done with the sole intent of creating a museum object.

Following a correspondence with the curator of the museum, we obtained a photocopy of two documents from the Archives of Pauls Stradins Museum of History of Medicine. The first document (in Russian) is an "Acceptance certificate" issued on 7 July 1966 regarding the acceptance of the taxidermied dog(s) made by Demikov in the Sklifosovsky institute. The document certifies that the dog(s) is(are) a suitable exhibit for the museum and is signed by the taxidermist (Danilov), by Demikov himself and by other members of museum staff. The other document (in Latvian) mentions that the dog(s) were received for free, they were sent to Riga by train, but the museum had to pay the taxidermist for his work (135.30 Rbl, which as we were told by the curator is about a month's salary of a worker or a lab assistant).

When we asked whether the dog(s) specimen was created purposely for a museum display, museum staff replied:

> As I learned – it is actually unknown, there are no documents preserved that the museum commissioned the creation of the dogs. It's folklore. I gathered the information from the 3 older colleagues, if they remember anything. Two of them think that the dogs were created, because museum wanted to have them for the exhibition purposes. But one of the colleagues is very convinced that this was not so – he told me that former museum director Karlis Arons (back then he wasn't director yet) went to Sklifosovsky and saw the dogs that were already created for experimental purposes. He was a dog lover and (was) in horror because of the experiment. According to my colleague – Arons told him that the small dog was chewing the ear of the larger one (and one ear of the large dog is actually damaged.) And then Arons talked with Demikov and the

> dog must be given to a taxidermist and then sent to Riga. The other two are very sceptical about this story – they have never heard that before, despite one of them was in a close relationship with Arons.
>
> (from authors' private correspondence with the Pauls Stradins Museum of History of Medicine)

Another, more recent example is that of an *ear-mouse* (i.e. a nude mouse with a tissue engineered human ear growing from its back) created by Chinese plastic surgeon and scientist Yilin Cao specifically for the opening of the Shanghai Science and Technology Museum. According to the curator there, the mouse was alive and present for the opening and was then preserved by plastination for permanent display. Professor Cao was part of the original team that created the first *ear-mouse* to be presented to the public during a 1995 episode of the popular BBC science television show, *Tomorrow's World*, dedicated to new breakthroughs in tissue engineering. In this TV documentary the presenter pauses to address the viewer, warning us that: 'you might find this next piece of film a bit disturbing, but we think it is such a significant medical advance that you really ought to see'. With this caveat, the BBC unveiled footage of the *ear-mouse*: a hairless mouse with a human shaped ear growing from its back, alive and displayed on a glass petri dish. The public and media response to this image was overwhelming, with the mouse and the scientists behind it receiving overnight notoriety. This led to one of the scientists – the same one that had decided to present the *ear-mouse* to the BBC against the advice of his colleagues – to cut the ear off the back of that mouse and cast it in a clear resin block as a museum ready object. He then approached the Smithsonian Institution and offered to sell it to the museum, which refused to acquire it. The infamous ear is still sitting on his office shelf waiting for a museum to purchase it.[4]

Conversely, we note artists working in the area of biotechnology embroiled in controversies concerning issues of authorship and intellectual property embedded in living artefacts which were created by them and/or created in a scientific laboratory for the sole purpose of making an art object. The art piece titled GFP *Bunny* by Eduardo Kac (2000), materialised in art galleries only in forms of traditional representations, such as photographs (potentially manipulated), ceramic sculptures or drawings. Yet, thousands of transgenic bunnies, some with green fluorescent protein (GFP) genes, are being created as part of biotechnological researches and applications. Could an artist identify one of them as an art object? Is it the context in which this bunny is presented (i.e. a performance in an art gallery) that turns it into an art object? Or is it the intention with which the rabbit was created – for an art display rather than as a scientific "tool"? Would the latter mean that techno-scientists are the new avant-garde?

The 19-century museum was opened to the public, to show the order of the universe, as revealed by science. Visitors could imagine themselves as participants in this new articulation of knowledge and power, they were invited 'To identify with power, to see it as . . . a force regulated . . . by society's ruling groups but for the good of all: this was the rhetoric of power embodied in the exhibitionary complex – a power made manifest not in its ability to inflict pain but by its ability to organize . . . an order of things and to produce a place for the people in relation to that order' (Bennett, 1995, p. 67).

By contrast, the practice of science since the late 19th century (following Jacques Loeb's view as science as engineering pursuit[5]) is shifting from observation of the living world to its manipulation through engineering. Museums seem to be following these trends by moving and showing biological specimens in the context of science and engineering sections of the museum – following the ethos of progress, innovation and public imagination. Aesthetic decisions are employed to consider these living animals as artefacts that represent human mastery over "nature".

Biological arts in natural history and science museum

Oxford dictionary refers to a museum as 'A building in which objects of historical, scientific, artistic, or cultural interest are stored and exhibited'. But how are intentionally artistic living artefacts presented within natural history and science museums?

Contemporary artistic biological artefacts are not commonly shown in the context of natural history and science museums. In 1998 one of our first artworks (a 2-dimensional photograph) became a background for a display of fossils as part of the *Diamonds to Dinosaurs* exhibition at the Western Australian Museum. However, since then the field of Biological Arts (*aka* Bioart) has garnered institutional acknowledgement within academia and, to some extent, within the broader art world.

Yet, in the context of science museums, biological art is predominantly presented through items that belong to the world of design, and which are thus displayed either in "innovation" areas or as a way to demonstrate contemporary techniques concerning the manipulation of life. The art, if at all, is contextualised in terms of scientific breakthroughs, current or future consumer products and almost always as a tool for traditional science communication.

The Center for PostNatural History (established by the artist Rich Pell) is a refreshing example of combining the ethos as well as aesthetics of the Natural History Museum. It sees itself as a place to 'study the origins, habitats, and evolution of organisms that have been intentionally and heritably altered by humans'.[6] In the tradition of natural history museums, it observes the phenotype, habitat and behaviour of the organisms on display rather than focusing solely on the technological mechanisms that took part in its evolution; it contextualises the organism within its novel ecology and umwelt.

Biological arts in art museums

While natural history museums are sometimes viewed as "dead zoos" they at least have some experience in working with biological specimens (and even have pests uniquely adapted to this environment such as the museum beetle – *Anthrenus museorum*) (Browne, 2015, p. 211).

The art museums however can be seen as the ultimate necropolis. The notion that the art object is eternal, never changing and commodified is still rooted within the art museum ethos.

Biological artworks embody the complete opposite. In this light, the reluctance of art museums to present biological artwork is hardly surprising. However, in recent years, a growing number of art institutes have opened up to presenting such art, starting with festivals such as Ars Electronica in Linz, then art and technology institutes such as ZKM, contemporary art galleries and finally more traditional art museums.

This shift also corresponds to the emerging recognition that biological technologies can be exploited and that these may bring profound changes in the public sphere.

Both artists and art institutions are engaged in a paradoxical dance of critique and celebration of the new found powers over life, driven and responding largely to the neoliberal innovation paradigm. Biological art artefacts are present manipulated life forms, signs of both hope and anxiety.

The scope of this chapter does not allow for an extensive survey of such instances however we will use some anecdotal case studies that point to *Neolife*-like tendencies and the erosion of the artistic display as a non-product.

Our art piece titled *Victimless Leather – A Prototype of Stitch-less Jacket grown in a Technoscientific "Body"* which was originally commissioned for a fashion show at the *John Curtin Gallery* in Western Australia (2004) was later presented at the Museum of Modern Art in New York in 2008, as part of the *Design and the Elastic Mind* Exhibition in the Design wing of the Museum:

> *Design and the Elastic Mind* explores the reciprocal relationship between science and design in the contemporary world by bringing together design objects and concepts that marry the most advanced scientific research with attentive consideration of human limitations, habits, and aspirations. The exhibition highlights designers' ability to grasp momentous changes in technology, science, and history – changes that demand or reflect major adjustments in human behaviour – and translate them into objects that people can actually understand and use.
> (Aldersey-Williams *et al.*, 2008)

The only piece to have living components as part of the show, *Victimless Leather* contained growing embryonic mouse cells. Other exhibits in the show that comprised living elements (plants, algae, etc.) had to resort to non-living imitation (green coloured liquid, plastic replicas of plants and so on). To our knowledge, *Victimless Leather* was the first piece to include living non-human material since Edward Steichen's hybrid delphiniums were exhibited as fine art in the Museum of Modern Art New York in 1936. What made our piece successful in this context, and closely aligned with our intent, was its spectacular failure. The piece ran into trouble a few weeks after the opening as *Victimless Leather*'s living cells grew into an embryonic body, clogging its life support system, resulting in the work being switched off. This "failure" not only raised public discussion regarding life and death in the context of living biological arts and beyond (Schwartz, 2008), it also acted as a counter balance to the celebratory context of the exhibition, offering less optimistic future scenarios.

Collecting Neolife

In another project, we explored the taxonomical crisis initiated by the advancements in biotechnology and the rise of the Neolifism. The piece *Odd Neolifism* (2010) was presented in art galleries including the Gallery of Modern Art in Brisbane Australia, that commissioned it for its exhibition *21st Century: Art in the First Decade*. In this work we poetically examined how historical knowledge can lead to shifts in the perception of life and how biology was becoming an engineering pursuit. The piece poses a question: where are tissue constructs – adapted to a laboratory ecology – positioned within taxonomical systems? In 1991, two evolutionary scientists, Van Velen and Maiorana, suggested that HeLa cells might constitute a new life form declaring that HeLa cells are the best-known cultured cells of human origin and that they have become a separate species restricted to a particular environment (Van Valen & Maiorana, 1991). In other words, HeLa cells could be considered an "odd" life form in a laboratory environment. In tissue banks that provide scientists with cell lines, one starts to find all sorts of oddities: for example cells that originate from up to three different organisms, such as fused human and mouse cells known as hybridomas. These cells are only classified by catalogue numbers or by very odd names. They are "Neolife". What place do they have in museums, which traditionally explain the taxonomy of life by showing us the living kingdom as a continuous developmental lineage, culminating with primates, the apes and Homo sapiens (often presented as odd artefacts of nature)?

In 2007 we developed an art work titled *NoArk*. Through these new, totally abstracted life forms – cells – the metaphor of the ark no longer applies. We see a return of those cabinets of curiosities, but this time, specimens come from supply catalogues for scientists and laboratories. Since we developed *NoArk*, an increasing number of museums have begun to collect fragments of life; frozen cells that represent the whole. These are a new addition to stuffed animals and specimens in jars, privileging the information represented by DNA over the phonotypical form of the organism. Here, the technology of collection is converging with the technology of making. New life forms are entering collections, yet collections are not complete . . . the "odd neolife" is not there in our natural history collections; lab-grown and modified life forms are still absent.

Odd Neolifism (2010) is an updated cabinet of curiosities. On one side of the work we placed a stuffed two-headed bird recalling the fascination with oddities and their significance in the 17th century. Museums have conventions in displaying preserved life forms, and in this display, ideas about a progressive complexity of species are questioned. At the far end of the display we included a living element – hybridoma cells growing within a bioreactor (a surrogate, technological body). These life forms are abstracted from their source, and yet they are living. Perhaps it is time to realise we need to find a place for Neolife.

Biomess: questioning our notion of life

In 2017, we were invited to collaborate with both the Art Gallery of Western Australia and the Western Australian Museum for the *Biomess* exhibition. This invitation prompted us to continue probing the strangest of living systems, both unintentional ("natural") and constructed ("designed") and to re-imagine how we can artistically convey the complexities of our own relationship with life in this time in history.

We live in a time when technology is becoming more lifelike, while life is becoming more technology-like; as our society allows technology more autonomy (i.e. driverless cars and artificial intelligence systems), we attempt to assert our control over living systems that existed independently from humans. With advances in genetic engineering and synthetic biology, living beings become more akin to cultural artefacts than to natural specimens. Our concept of life is assaulted when life is treated as a technology and/or merged with technology, both wet and dry. As life becomes a raw material for human desires, constructed life escapes science labs to become a medium for artistic expression and consumer products. New life forms that defy classification begin to appear while concurrently we still do not understand many existing lifeforms – ultimately leading us to question our traditional notions of identity, self and bodies. *Biomess* celebrates and challenges the strangeness of life by using luxury retail aesthetics to make non-charismatic life forms into objects of desire. Living organisms, natural history specimens and lab-grown life, are combined and presented with "bio-Gucci" shop aesthetics.

Following philosopher Timothy Morton, we play with the notion that Nature 'is normative, which is a philosophical term for something that establishes differences between the normal and the abnormal, often with ethical overtones' (Morton, 2016, p. 146). We are exploring life forms that are "un-normative" and seem to defy common ideas about self, gender, identity and individuality as well as liveness, artificiality and technology. This installation presents living organisms, preserved/taxidermised specimens alongside technologically "designed" and manipulated life forms (cell lines, tissue constructs and hybridoma cells, growing and changing in customised environments). We are asking what kind of aesthetics should be used when life becomes a commodity or object of desire. How do we make sense of living forms that were constructed to blur what we consider specimen, scientific tool, product and art object?

If museums are a *normalising place*, or as Bennett suggests a disciplining place of society's evolving morals (Bennet, 1995), then we present some of these human-made monstrosities as tools and objects for the myth of progress but also stand for pleasure, entertainment and artistic articulation. Morton's suggestion that *Nature* is the monster, informs our stance that knowledge in the *Museum* is a dominion in flux when dealing with life that is manipulated by humans.

Notes

1 The authors refers to such approach as *Neolifism* (Catts & Zurr, 2016).
2 German word meaning "environment" or "surroundings", firstly used by Jakob von Uexküll (1909).
3 See more about Christiaan Barnard (Encyclopædia Britannica, 2018).
4 Based on recoded interviews and email exchange between Oron Catts (Author) and Prof. Charles Vacanti, 2014.
5 For more see Pauly, 1987 or Catts and Zurr, 2014.
6 See more about The Center for PostNatural History (Center for PostNatural History, n.d.).

Bibliography

Aldersey-Williams, H., Hall, P., Sargent, T. & Antonelli, P. (2008) *Design and the Elastic Mind*, New York: The Museum of Modern Art.

Bennett, T. (1995) *The Birth of the Museum: History, Theory, Politics*, London: Routledge.

Browne, M. (2015) *Practical Taxidermy – A Manual of Instruction to the Amateur in Collecting, Preserving, and Setting Up Natural History Specimens of All Kinds*, Redditch: Read Books Limited.

Cameron, F. (2008) Safe Places for Unsafe Ideas? History and Science Museums, Hot Topics and Moral Predicaments, in: Terwey, M. (Ed.) *Social History in Museums: Journal of the Social History Curators Group*, 32, 5–16.

Catts, O. & Zurr, I. (2018) Artists Working with Life (Sciences) in Contestable Settings, *Interdisciplinary Science Reviews* [Online], 43(1), 40–53. Available at: www.tandfonline.com/doi/full/10.1080/03080188.2018.1418122 (Accessed 11 May 2018).

Catts, O. & Zurr, I. (2016) The Biopolitics of Life Removed from Context – Neolifism, in: Wilmer, S.E. & Zukauskaite, A. (Eds.) *Resisting Biopolitics*, New York and London: Routledge.

Catts, O. & Zurr, I. (2014) Countering the Engineering Mindset – The Conflict of Art and Synthetic Biology, in: Calvert, J. & Ginsberg, D. (Eds.) *Synthetic Aesthetics*, Boston, MA: MIT Press.

Center for PostNatural History. (n.d.) *About* [Online]. Available at: www.postnatural.org/About (Accessed 09 May 2018).

Encyclopædia Britannica. (2018) *Christiaan Barnard* [Online]. Available at: www.britannica.com/biography/Christiaan-Barnard (Accessed 09 May 2018).

Morton, T. (2016) Frankenstein and Ecocriticism, in: Smith, A. (Ed.) *The Cambridge Companion to 'Frankenstein'*, Cambridge: Cambridge University Press.

National Museums Scotland. (n.d.) *Dolly the Sheep* [Online]. Available at: www.nms.ac.uk/explore-our-collections/stories/natural-world/dolly-the-sheep/ (Accessed 09 May 2018).

Pauly, P.J. (1987) *Controlling Life: Jacques Loeb & the Engineering Ideal in Biology*, New York: Oxford University Press.

Rose, H. & Rose, S. (2007) The Wonders of the Recurated Wellcome Collection, *The Lancet*, 370(9582), 125–126.

Schwartz, J. (2008) Museum Kills Live Exhibit, *The New York Times* [Online], 13 May. Available at: www.nytimes.com/2008/05/13/science/13coat.html (Accessed 09 May 2018).

Science Gallery International. (n.d.) *The Global Science Gallery Network* [Online]. Available at: https://international.sciencegallery.com/network (Accessed 11 May 2018).

Uexküll, J. von. (1909) *Umwelt und Innenwelt der Tiere*, Berlin: J. Springer.

Van Valen, L.M. & Maiorana, V.C. (1991) HeLa a New Microbial Species, *Evolutionary Theory*, 10(2), 71–74.

Vermeulen, N., Tamminen, S. & Webster, A. (Eds.) (2016) *Bio-Objects: Life in the 21st Century*, London: Routledge.

6.3 Decolonising natural history museums through contemporary art

Bergit Arends

How do we acknowledge the moral values and the violence of past collecting practices in the natural sciences? How do we reflect on these practices now? Natural history museums, like other museums, are dynamic, cultural formations shaped by society (Findlen, 1994; Jardine *et al.*, 1996). Historically and in the present time artists and scientists are both part of cultures that seek to discover and to represent the natural world. Natural history museums thus have become complex archives of the representations of nature and they are places in which to acknowledge and to reflect. A contemporary artist, who spent time studying historic collections in the Natural History Museum in London, reflects on the late 18th and 19th century formations of museums in stating: 'The main thing for me was the science aspect [. . .] it all relates to each other, adds together: history; the Enlightenment; colonisation' (artist quoted in Porter, 2012, p. 6). These centuries were characterised by European fantasies of knowledge made into power (Richards, 1993) and by imperial authority over the non-Western world and over non-Western objects (Barringer & Flynn, 1998). These museums, which originated in the Western world, are now under scrutiny from diverse cultural perspectives.

Hybridising museological, artistic and historic scientific practices

In the hereby described works, I adapt 'decolonising' to a critical museological process in order to productively interrogate the principles of natural history museums. The term 'decolonisation' was coined in the 1930s by German economist Moritz Julius Bonn (1873–1965). In his work on imperialism he refers to the processes through which colonial rule dissolves. Bonn also used the term 'counter-colonisation' as synonym for decolonisation. Both terms refer to the ending of colonial rule as well as the developments leading to the demise of colonial empires. Studies of decolonisation look at transfers of power, the diversity of languages and the hybridisation of cultures. Significantly, these processes link the history of Europe with the histories of other continents right up to the present time (Rothermund, 2006). In a museum context 'decolonising' can mean 'resisting the reproduction of colonial taxonomies, while simultaneously vindicating radical multiplicity' (L'Internationale Online, 2015, p. 5).

The relationships of museums to places, local and global, are expressed through their collections, research infrastructures, and the diversity of audiences. For this specific study I am inspired by advances made in ethnographic museums in working with artists. Ethnographic museums offer perspectives and collections that are outside of European scientific practices and the ordering of the natural world. They draw our attention to the different relations other cultures have to the natural world. These collections and ways of working with representatives of other cultures broaden our understanding of what constitutes the study of the natural world.

Therefore, the discussion of this enquiry is set into a current museological context of notable examples of museums engaging with artists' voices reflecting on how humans live within the natural world. These are projects such as the Humboldt Lab Dahlem, Germany (2012–2015). The Lab contributed to the exhibition planning of the Humboldt Forum, due to open in the centre of Berlin in 2019. The Humboldt Forum as a museum space seeks to present the universal themes and global developments of humans through the non-European collections of the National Museums in Berlin. Significantly the project explores origins, ownership, and methods of presentations of these collections, but the project's origins and genealogy have been marred from the outset by controversies surrounding the colonial origins of the collection and the museum building.

A diversity of Western artists' perspectives onto human and non-human species relations was explored in the Art/Nature Programme at the Natural History Museum, also in Berlin (2014–2018). The itinerant project 'Indigeneity in the Contemporary World: Performance, Politics, Belonging' (2009–2014) initiated, facilitated and documented artists' residencies to investigate how indigeneity is understood in a globalising world at the Horniman Museum in London, which holds anthropological and natural history objects, and the Berlin Ethnological Museum, whose collection is one to be moved into the Humboldt Forum (Gilbert, n.d.).

The Museum of Archaeology and Anthropology in Cambridge, United Kingdom, is another example in engaging contemporary artists in the field of ethnography through which a diversity of engagements with land can be observed. The Wellcome collections of medical-ethnographic objects draw on a range of explorations into nature through the lenses of human and animal wellbeing. Historic country houses are often sites of both natural and ethnographic collections, gathered by former owners. In the UK these houses and their gardens are now often cared for by the National Trust. Artists, questioning taxonomy and nature as resource for profit, have worked with collections of botanic gardens, such as the economic botany collection and the public spaces of the Royal Botanic Gardens, Kew (Arends, 2018). To address artists' decolonising practices of natural history museums I have made a methodological sideways step by considering the challenges of ethnographic museums and the relevance of place as well as national and European histories.

Challenging Western natural history museum practices

How do 21st-century contemporary artists from non-Western cultures engage with natural history museums? Contemporary artists can offer conceptual and visual methodologies that reveal the complexity of producing knowledge and can question the underlying structures and motivations. Artists are agents of cultural knowledge exchanges and contribute to the growing significance of research into collections and the search for new forms of representation and dialogue. Artists can thus contribute to museums as dynamic institutions whose cultural and moral substance needs constant assessment in a contemporary global world. Here, I use the Natural History Museum in London as an example to show how an institution of the British Empire became and remains connected to Australia, China and India.

The Natural History Museum in London has its origins in the British Museum. The museum's founding collections from 1753 were donated by physician Sir Hans Sloane (1660–1753) and encompass natural history, ethnographic objects, antiquities, coins, artworks, books, and manuscripts. His collections became dispersed between the British Museum, the British Library, and the Natural History Museum. As the natural history department outgrew its original Bloomsbury site, a purpose-built museum housed the collections and its curators in South Kensington from 1881 onwards. Some of Sloane's objects collected in the 18th century in the Atlantic are associated with slavery (Delbourgo, n.d.).[1]

Natural history museums relate to places through the gathering of knowledge about the natural world in a variety of forms and data, such as material objects, textual representations, and visualisations. The museum collects according to a system, its principles reflect the organisation of the natural world through scientific, economic and aesthetic ideas. Typically, the system is reliant on a global network of scientific collaborators and collectors, out of which specific relationships to places have emerged. Scientific research, forms of storage and display encapsulate these myriad European relations with the natural world. European trade routes and nascent forms of global capitalism, mapping the world and modern public scientific culture build upon each other. For example, English naturalist and apothecary James Petiver (c. 1663–1718), whose vast and eclectic insect collections from the East Indies, the Caribbean and China are now held at the Natural History Museum in London, were gathered by a network of collectors. Petiver relied on the paid assistance of traders and mariners, the shipping network of the slave trade in the British Atlantic and the collections of Caribbean insect specimens made with the free labour from black slaves. Similarly, Maria Sybilla Merian (1647–1717), the naturalist and illustrator of stylised representations of insects, relied on indigenous labour and slaves to assist with her entomological work in the Dutch colony of Suriname (Gandy, 2016).

Artists working with historic, scientific illustration collections

At the museum there is plenty of potential to work on troubled pasts and presents, on social and disciplinary hierarchies and ongoing discussions on restitution. We wanted to reconsider museum practices and scientific objects. By providing a display space we engaged in the politics of representation. We could experiment with the juxtaposition of historic works and contemporary artworks. How would different styles of interpretation sit next to each other? How complementary, how uncomfortable, how obscure would a change of aesthetics and visual language be for our audiences – and for museum staff? In brief, the institution did seek to challenge itself in working with artists and to provide a public space for these challenges.

The case studies to follow are based on my work at the Natural History Museum London, where I was Curator of Contemporary Art from 2005 to 2013, during which I commissioned artists to engage with the museum's collections, scientific research and the role of the museum in today's society. By working specifically with international artists and by interrogating the historic scientific illustration collections we developed a decolonising approach to the museum's work. Artists were invited to take up residency in London as part of the International Artist Research Residence Programme at the Natural History Museum in partnership with the artists' studio organisation Gasworks, London.

We hosted three artists: Shanghai-based artist Hu Yun (in 2010), Aboriginal Australian painter Daniel Boyd from Sydney (in 2011), and Bangalore-based artist Sunoj D (in 2012). They each spent a three-month research period at the museum and developed works for an annual display in its then newly opened Images of Nature Gallery. The gallery offered the possibility to integrate the artists' commissions from its inception. This gallery, unlike most of the other exhibition spaces, is aimed at an adult audience and describes historic and contemporary practices of natural science from observation to recording to communication. Most importantly for the art programme, the space combined permanent with temporary, thematic displays. The artists' commissions became integrated into the displays, which in the first three years explored historic scientific illustrations from specific geopolitical areas. Hu Yun started with the John Reeves collection from the early 19th century, Daniel Boyd with the drawings and watercolours from the so-called First Fleet, made when the first British colonisers settled in Australia, and Sunoj D from the numerous collections relating to Indian natural history. Through the respective artists' works I explain the different artists' strategies of de-colonising the Natural History Museum.

Hu Yun: commissioning a collection

Hu Yun studied the more than 2,000 botanical and zoological drawings collected by John Reeves (1774–1856), who was based in Canton (present-day Guangzhou) and Macao between 1812 and 1831.[2] Reeves, an East India Company tea inspector and naturalist, is by the museum considered to be the most important British naturalist in China during the economic boom before the First Opium War.[3]

Hu Yun became interested in the production of this illustration collection by studying the relationship between its commissioner, Reeves, and the Chinese artists-draughtsmen instructed by him. Hu Yun's work, titled *Memorialise Memory*, became a site-specific installation which adapted to the aesthetic language of the Edwardian wooden display case by using a similar wood for his work. The text inscription chiselled into his wooden inserts related the life and scientific work of Reeves and adopted a language of a commemorative plaque. The commemoration for the Chinese illustrators, referred to as Akut, Akam, Akew and Asung in Reeves' ledger book, is a continuous video running in reverse, projected onto the wooden surface inserted into the display case. Hu Yun filmed himself writing their names, which are then subsequently erased, 'memorialised' and re-written. The four draughtsmen – we expect there were more than these four to illustrate specimens – were written out of history through scientific conventions. At the time collections tended to be named after the commissioner. Contributors to collections, such as the commissioned illustrators, were usually not named as authors or co-creators. Hu Yun himself anonymised his installation in the gallery space and the separate drawings on paper, which he donated to the museum's library collection, by omitting his name. His work exudes wit and humour, jarring historic scientific conventions with those of contemporary art in which authorship is key. His research methodology was that of an archivist-historian of science and technology as adopted by artists. Art historian Foster (2004) describes a range of motivations driving artists to 'probe' the archive: to 'physically' present (often) lost or displaced historical information, alternative knowledge or counter-memory; to de-familiarise or disturb the legibility of current knowledge, and to push originality and authorship. Hu Yun adopted these strategies.

Sunoj D: co-creating a collection

The theme of authorship in networks of co-creation was also pursued in the work by Sunoj D. The artist, born in Kerala, focused on the Indian art collections and the *Hortus Malabaricus* (Garden of Malabar). Commissioned by the Dutch East India Company and published between 1678 and 1703, this plant record is a collaboration of more than 100 contributors from different professions and castes. Contributors included Brahmin physicians Ranga Bhat, Vinayaka Pandit, Appu Bhat, together with the Malayali physician Itti Achuden, a doctor from the Ezhava community, to document 742 medicinal plants found in Malabar (now Kerala). Sunoj's engagement with the historic works was participatory and personal: Sunoj had brought spices from the Malabar area, where his family lives, and created social gatherings. He cooked Indian dishes, using his spices, engaged people in discussion, and asked them to contribute flowerpots from their homes and gardens to the installation at the museum. This part of the project responded to the artist's interest in our understanding of where nature can be found and encountered. The participatory project involved museum staff as well as community groups from whom plant pots were donated – a departure from the museum's exhibition practices, yet an interaction strategy somewhat equivalent to the scientific research activities offered in the museum's Angela Marmont Centre for UK Biodiversity. The centre supports individuals, schemes and societies that record, monitor and protect the UK's biodiversity and encourages interactions with staff through its identification services. Equally, numerous citizen science projects at the museum enable participation in the doing of science and observation of environmental changes.

The work Sunoj created in the gallery space also included a paper scroll. Sunoj D's scroll drawing *Somewhere Between Living and Dying* (2013) assembles plants from the Hortus and depicts them in wild disorder, a stark contrast to the systematic principles of the museum's later plant collections. Different types of formats for presenting herbaria, reflecting both changing cultural fashions as well as scientific progress, can be observed in the historic herbaria, such as the seven bound volumes of plant specimens in the Sir Hans Sloane Herbarium, collected on Sloane's voyage to Jamaica (1687–1689) or the 3,000 plant specimen herbarium of wealthy Anglo-Dutch merchant George Clifford (1685–1760), all of which were influenced by the cultural aesthetics of their time.

Daniel Boyd: the currency of colonial images

Daniel Boyd spent time researching the First Fleet Collection, which consists of 629 drawings and watercolours. The records were done by Thomas Watling, a convicted forger, and George Raper, a midshipman of the fleet's lead ship, HMS Sirius, as well as a number of unknown draughtsmen, referred to collectively as Port Jackson painters. Created from 1788 onwards the watercolours, ink and pencil drawings include maps, landscape views, flora and fauna, portraits of individuals of the Eora Nation. Eora refers to the group of clans, united by a common language, that lived in the area around what was Port Jackson, as referred to by the settlers, or Warrane by the indigenous population, now modern-day Sydney. The collection also includes depictions of how the indigenous peoples inhabited the landscape and scenes of conflict with the British settlers. The artist's initial interest was to probe how colonial images inform contemporary society and how we interact with these images today. For example, in relation to one of his earlier works Boyd reflects on the 1902 history painting *The*

Landing of Captain Cook at Botany Bay, 1770 by E. Phillip Fox: 'Shifting the proposed view of Fox's painting to something that was an Indigenous person's perspective allowed me to challenge the subjective history that has been created' (Boyd, 2008). Boyd 'counter-colonised' Fox's work with his own painting *We Call them Pirates Out Here*, subverting and countering Fox's iconography by replacing the well-known cast of European characters who set foot on this land – Captain Cook, the wealthy adventurer and later president of the Royal Society Sir Joseph Banks, botanist Daniel Solander – with personal and indigenous people. Boyd often reworks archival imagery and history paintings made by Europeans to construct a history of Australia from 'out here', signalling a contemporary Aboriginal standpoint (Nugent, 2009). In his gallery installation *Up in Smoke Tour* (2011) Boyd displayed a series of watercolours held within archival boxes, like those used in the museum's anthropology collection now. His images are appropriations of historic images of Sydney harbour, plants, a human skull from Australia, a portrait of an individual of the Eora Nation, and sculptures from Vanuatu. His layered watercolours deliberately fractured the historic image. These works are 'like an erasure of memory or history in an image or an object [. . .] The loss of information empowers *me* – the viewer is put in the position where they don't have that information' (quoted in Porter, 2012, p. 8).

Following the analytical activity of the 'artist-as-archivist', here the artist developed strategies to challenge the representation of history (Godfrey, 2007). Boyd's body of watercolours, made for the commission at the Natural History Museum, addressed national histories and natural histories. He deconstructed ways of seeing, ideas about land, colonial relations and encounters. This process allows Boyd to relate dispossession and colonisation to his own contemporary experience as an Aboriginal from the peoples of Kudjla, Eastern Kuku Yalanji, Kangulu, Jagara, Bandjalung, Kuku Djungan and the North Pentecost Island (Vanuatu) peoples and to reclaim his own history.

Further to his initial research Boyd's interest in the museum's anthropology collection led him to investigate possible ancestral links to individuals in the human remains collection and how the Museum deals with remains currently in the museum's possession. His search led him to a skull believed to be taken from a burial ground west of Brisbane.

Conclusions

The artists' scholarship was not dispassionate, but intensely interested and had extraordinary attentiveness to 'evidence' (Thomas & Losche, 1999). The methodological approaches were those of appropriation of materials, museum conventions and science. The artists employed historical and archival research methods in order to draw attention to gaps in and ownership of knowledge. Furthermore, in studying specific material objects through their biographies, the artists addressed questions of ownership that might lead to restitution. The museum became a 'contact zone' for the hybridisation of visual, material, performative and language cultures in which collection stores and archives became a 'site of translation' (Clifford, 1997). The artists' works and the processes leading up to these showed that for these artists, 'science' is a set of cultural practices through which knowledge is constructed encompassing collecting practices, modes of representation within their historical and colonial or imperial contexts. This realisation affected staff members and the public alike. A museum staff member responded to the artists' presence in the museum: 'They have come in as artists *and* non-Europeans, with a very different background and perspective – on colonialism, on the European view of science [. . .] a very different take which generated discussion and raised questions not raised before' (Porter, 2012, p. 6).

The representation of issues was challenging. Again a museum staff member reflected on this translation: '[Hu Yun's] was "invisible artwork" – you could really easily miss it! This was his intention but others didn't 'get it'. He didn't want his name to be mentioned [. . .] he played on the conventions. Staff understood the programme would be happening and expected something show-stopping, dazzling, big. Did his subtlety and sophistication mean that they valued the work less? It was delightful, playful *and* serious; disturbing preconceptions' [emphasis in original] (Porter, 2012, p. 7). A visitor evaluation commissioned in the first few months of the gallery's opening has a visitor comment on the display: 'You have got some quite traditional work and then something like that which is more of an installation and more modern, so you are really crossing all kinds of cultures' (Morris *et al.*, 2011, p. 26).

In this respect, specifically the collections and archives of natural history museums can also contribute to the search for new forms of collaborative research in creating opportunities for dialogue with indigenous peoples (Cameron, 2014), by drawing attention to the gaps in collections and collecting practices (Anderson, 2015; Driver & Jones, 2009; Stoler, 2009) and in finding diverse terms of engagement with the natural world towards both environmental justice and human rights (Demos, 2016).

Here the artists contributed to how we re-think who archives and collections are for, how they can be opened up and whose histories we tell (Edwards & Mead, 2013; Harrison *et al.*, 2013).

By de-colonising I referred to revisiting and challenging the institutional taxonomies, scientific practices and colonial impositions while speaking out for cultural multiplicity and recognition. The artists' works represent complex issues in a new, deeply personal language, giving agency to objects and creating new social relations. The institution opened up to give representational space to artists as agents, reflecting on the history of the natural sciences, but this exciting initiative was cut short when the museum later decided that cultural practices had no place within a scientific research institution. Thus frustratingly, it remains in the power of the institution to acknowledge, to reflect, and to change. Though programmes such as this one as well as initiatives like Museum Detox, the network of BAME (Black, Asian and Minority Ethnic) heritage workers, are now influencing the UK museum sector and are challenging narratives, programming and are calling for decolonisation (Das & Lowe, 2018). Institutional critique leading towards the decolonisation of a natural history museum is, I suggest, transferable to the sites of science and other museums or indeed other archives and collections-based institution in the European heritage sector. Wintle (2013) argues that material culture and museums not only reflect wider political change, but exercise agency on processes of decolonisation by acting as microcosms of political encounters. We need to learn how to acknowledge the benefits of such encounters. A positive disquiet arises out of the interchange between our looking at historic objects and for the object to look back at us. We need to learn how to feel the agency historic objects have over us now. Museums are cultural formations and are shaped by society. But we are in turn shaped by objects held in museums. Objects help us to transition from one historic context into another, from one collective system of emotions and orientations to another. There is positive disquiet enjoyed from the changes effected within oneself and to the changes in relation to others from different cultural system of beliefs and values (Savoy, 2018, p. 8), let alone those from the disciplines of art or science.

Finally, we should feel inspired to not simply address national, but European histories in a global context. Artists and scientists alike can create diverse fora for public, critical exchanges. These are to address injustices embedded in the practices of science, to find new forms of critique and to co-create new knowledge.

Notes

1 To address the Jamaican heritage of plant specimens in the Sloane Herbarium, specifically *Theobroma cocao* (chocolate), we worked with artist group Motiroti and a community of Jamaican elders as part of the *Priceless* festival in 2006. Together we related their individual stories about chocolate and its culinary uses.
2 See more about Indian and Chinese art collections at Natural History Museum (Natural History Museum, n.d.).
3 The First Opium War took place in 1839–42; the Second Opium War in 1856–60.

Bibliography

Anderson, D.M. (2015) Guilty Secrets: Deceit, Denial, and the Discovery of Kenya's 'Migrated Archive', *History Workshop Journal*, 80(1), 142–160.

Arends, B. (2018) The Subversive Potential of Seeds in Three Acts (2010, 2013, 2014), in: Zinnenburg Carroll, K. (Ed.) *Botanical Drift: Protagonists of the Invasive Herbarium*, 96–109, Berlin: Sternberg Press.

Barringer, T. & Flynn, T. (Eds.) (1998) *Colonialism and the Object: Empire, Material Culture and the Museum*, London: Routledge.

Boyd, D. (2008) *About the Artwork* [Online]. Available at: www.mca.com.au/collection/work/2006.25/ (Accessed 08 February 2018).

Cameron, L. (2014) Participation, Archival Activism and Learning to Learn, *Journal of Historical Geography*, 46, 99–101.

Clifford. (1997) *Routes: Travel and Translation in the Late Twentieth Century*, Cambridge, MA: Harvard University Press.

Das, S. & Lowe, M. (2018) Nature Read in Black and White: Decolonial Approaches to Interpreting Natural History Collections, *Journal of Natural Science Collections*, 6, 4–14.

Delbourgo, J. (n.d.) *Slavery in the Cabinet of Curiosities: Hans Sloane's Atlantic World* [Online]. Available at: http://britishmuseum.org/research/news/hans_sloanes_atlantic_world.aspx (Accessed 04 February 2018).

Demos, T.J. (2016) *Decolonizing Nature: Contemporary Art and the Politics of Ecology*, Berlin: Sternberg Press.

Driver, F. & Jones, L. (2009) *Hidden Histories of Exploration: Researching the RGS-IBG Collections*, London: Royal Holloway, University of London, in association with the Royal Geographical Society with IBG.

Edwards, E. & Mead, M. (2013) Absent Histories and Absent Images: Photographs, Museums and the Colonial Past, *Museum and Society*, 11(1), 19–38.

Findlen, P. (1994) *Possessing Nature: Museums, Collecting, and Scientific Culture in Early Modern Italy*, Berkeley and London: University of California Press.

Foster, H. (2004) An Archival Impulse, *October*, 110, 3–22.

Gandy, M. (2016) *Moth*, London: Reaktion Books.

Gilbert, H. (n.d.) *Indigeneity in the Contemporary World: Performance, Politics, Belonging (2009–2014)* [Online]. Available at: www.indigeneity.net/residencies/ (Accessed 11 April 2018).

Godfrey, M. (2007) The Artist as Historian, *October*, 120, 140–172.

Harrison, R., Byrne, S. & Clarke, A. (Eds.) (2013) *Reassembling the Collection: Ethnographic Museums and Indigenous Agency*, Santa Fe: School for Advanced Research Press.

Jardine, N., Secord, J.A. & Spary, E.C. (Eds.) (1996) *The Cultures of Natural History*, Cambridge: Cambridge University Press.

L'Internationale Online. (2015) *Decolonising Museums* [Online]. Available at: www.internationaleonline.org/media/files/decolonisingmuseums-2.pdf (Accessed 08 February 2018).

Museum Detox. (n.d.) Available at: https://museumdetox.com/ (Accessed 07 April 2019).

Morris, Hargreaves & McIntyre. (2011) *Natural Selection: An Evaluation of the Images of Nature Gallery at the Natural History Museum*. Unpublished evaluation report.

Natural History Museum. (n.d.) *India and China* [Online]. Available at: www.nhm.ac.uk/our-science/departments-and-staff/library-and-archives/collections/india-and-china.html (Accessed 24 June 2018).

Nugent, M. (2009) *Captain Cook Was Here*, Port Melbourne, Vic.; Cambridge: Cambridge University Press.

Porter, G. (2012) Evaluation of the International Residency Programme, in: *The Images of Nature Gallery at the Natural History Museum*. Unpublished evaluation report.

Richards, T. (1993) *The Imperial Archive: Knowledge and the Fantasy of Empire*, London: Verso.

Rothermund, D. (2006) *The Routledge Companion to Decolonization*, London and New York: Routledge.

Savoy, B. (2018) *Die Provenienz der Kultur*, Berlin: Matthes & Seitz.

Stoler, A.L. (2009) *Along the Archival Grain: Epistemic Anxieties and Colonial Common Sense*, Princeton, NJ: Princeton University Press.
Thomas, N. & Losche, D. (1999) *Double Vision: Art Histories and Colonial Vision in the South Pacific*, Cambridge: Cambridge University Press.
Wintle, C. (2013) Decolonising the Museum: The Case of the Imperial and Commonwealth Institutes, *Museum & Society*, 11(2), 185–201.

Further reading

Arends, B. (2015) Pointillist Observations in Notebooks of Earth and Universe, in: Rule, D. (Ed.) *Daniel Boyd: The Law of Closure*, 69–87, Melbourne: Perimeter Editions.

6.4 Case studies

6.4.1 Imagining the future

Many contemporary artists working on scientific topics tend to include sociological and ethical aspects in their works, doing precisely what Science Fiction does, i.e. imagine alternative futures and realities.

Today, the power of imagination in combination with hard facts and data is being taken into consideration not only in science engagement environments but also in more academic art settings. For example, the 2016 Biennale of Sydney highlighted relationships between contemporary art and Science Fiction in its 20th edition, titled "The future is already here – it's just not evenly distributed".

Case study:
Ars Electronica Futurelab, Linz (Austria)

Maria Pfeifer, Senior Curator and Key Researcher

Ars Electronica, since 1979, has sought out interlinkages and congruities, causes and effects bringing together art, technology, society. It circulates ideas that are innovative, radical, eccentric in the best sense of that term. It aims to influence our everyday life, our lifestyle, our way of life.

Beyond the Lab – The DIY Science Revolution

A travelling exhibition that presents seven exemplary stories of 'DIY scientists' from around the world and includes three original works created by artists who position themselves at the crossroads between art, science and technology. In the exhibition these art pieces are used as an integral chapter of the storytelling, to describe possible future visions and scenarios through speculative technological objects, artistic prototypes and a fictional documentary film.

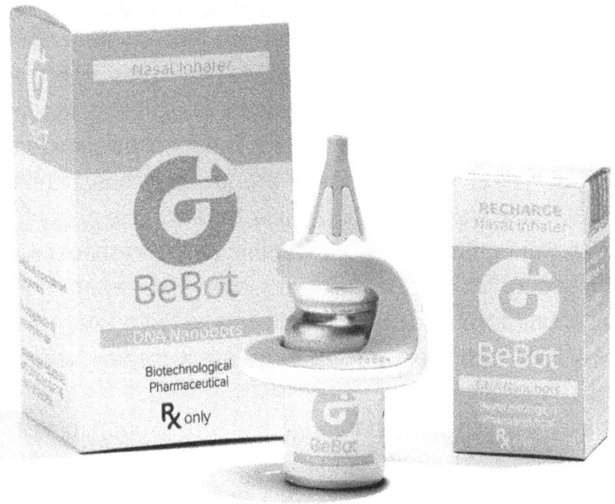

Figure 6.4.1 Artists Jacob and Illera have created the BeBot, an imaginary nanobot that interferes with our nervous system to suppress our conscious and subconscious dietary desires

Source: STAUDINGER+FRANKE

Description

The exhibition was developed in 2015 by the Science Museum in London for the Sparks project, funded by the H2020 European Commission program and involving 33 actors (science shops and science centres/museums) from 29 European countries.

The exhibition presented seven stories of 'DIY scientists' from across Europe comprising a diverse selection of objects, photography, film and contemporary art pieces. These stories revolved around three content themes: DIY Biology; Health Hacking; Citizen Science.

A fourth section presented three artworks selected by a jury composed by experts from Ars Electronica, Science Museum London, and the Amsterdam Health and Technology Institute.

The three artworks – by Anouk Wipprecht (Netherlands), Lucy McRae (UK), Jakob and Lea Illera (Austria) – were created during artist residencies at Ars Electronica Futurelab, where artists imagined where DIY science and medicine might lead us in the future and what this could mean for people's lives. During the residencies artists, researchers, producers and curators worked together on developing and producing the art pieces for the exhibition: Anouk Wipprecht built 3D-printed unicorn horn-shaped headsets (*Agent Unicorn)* designed for children with ADHD (attention deficit hyperactivity disorder) that log what strikes their attention and improve ADHD care. Lucy McRae produced the observational fictional documentary film *The Institute of Isolation*, exploring the idea of body optimisation through reactions to sensory deprivation and extreme isolation. Jakob and Lea Illera produced a speculative piece called *BeBots* – fictional nanorobots, able to block the appetite for unhealthy food, exploring further the relationship between robots and humans.

The exhibition also featured a small event space for activities and workshops (developed by Copernicus and WiLa Bonn) and a small 'local display' that host venues used to create a display based on locally relevant 'DIY science' research.

The exhibition travelled through Europe in four copies simultaneously, first opening in July 2016 at the Science Museum (London), Copernicus Science Centre (Warsaw), Science Shop (Bonn) and Hiša Eksperimentov (Ljubljana).

The primary target audience for the exhibition comprised ordinary museum visitors (i.e. not science specialists) aged 12 and over, with tours and programs delivered by local art and science explainers.

Outcomes

The overall aim of the project was to promote Responsible Research and Innovation (RRI) in Europe and show that there are ways to get directly involved in science and that the responsibility for scientific research and innovation is shared by various stakeholders.

The specific aim of the artist residencies was to produce compelling visual representations of future innovations that could provide a disruptive point of view about the social impact of science and research helping to raise general awareness on the subject of social responsibility and consequences.

Presenting artworks within a science exhibition – a very common approach at Ars Electronica – was a practice somewhat strange to many of the other hosting venues.

The key idea of including artworks together with science stories was to show artists' visualisations of possible future scenarios. And the key challenge was to present the artworks not as something different and separate but as an equal chapter of the science stories.

In general, to better understand the scientific aspects of the artworks and the artistic and aesthetic aspects of science stories, visitors were requested to visualise the links and connections between the art and the science. This operation was quite complex and the low level of interactivity in the exhibition did not help, which left the task of interpretation to local staff that had to proactively and constantly use exhibition content to stimulate discourse among visitors.

If it is true that art pieces can serve as mediators between the public and the world of scientific research, interpreting and questioning scientific approaches, it is also true that their presence can sometimes challenge curators in their effort to present a coherent narrative that fluctuates between fact and fiction. In this situation human mediation can become a key factor in the success of hybrid exhibitions.

6.4.2 Questioning the roots of science

We commonly think of science as a human practice that tends to be unbiased, universal, factual. The philosophy of science probes these assumptions by asking what can qualify as science and investigating the methods and purposes of science itself. In this sense the philosophy of science intersects with studies on epistemology, ontology and metaphysics, questioning the definition of "truth". Sociology also comes into the picture when researchers such as David Bloor investigate how knowledge is created from a sociological perspective, maintaining, for example, that all scientific theories are influenced by researchers' biases and other sociological factors.

Contemporary art is rooted in philosophy and very often dedicated to uncovering deeper layers of reality. Just as in philosophical research, there are several examples of artistic practices that question the very methodological and ontological foundations of science and of the related concept of "truth".

Case study: CosmoCaixa, Barcelona (Spain)

Joan Foncuberta, Essayist and Photographer

CosmoCaixa occupies the premises of what was the first interactive Science Museum in Spain, inaugurated in 1981. Now the new CosmoCaixa aims to be a meeting point between science and society and provides more than 30,000 square metres of facilities devoted to promoting science.

The Sputnik: Soyuz 2 Odyssey

An exhibition including cultural objects from the history of science and technology alongside fictional photographs and artefacts. The artistic work created a deceptive environment that gave the opportunity to highlight the importance of the scientific method as a tool to unmask deception and question the "truth".

Description

The *Sputnik: Soyuz 2 Odyssey* exhibition consists of a hundred photos, graphics, maps, uniforms and a replica of the Soyuz 2. The individual objects establish a dialogue between the two interconnected worlds of art and science and make sense when viewed as a whole.

The exhibition presents one of the unsolved mysteries of the history of the conquest of Space: the disappearance of cosmonaut Ivan Istochnikov mid-flight, victim of the failed Soyuz 2 expedition.

The story began in 1957, with the successful launch of the first artificial satellite, the Sputnik 1. This event marked the beginning of the conquest of Space but also of an obsessive rivalry – in the frame of the cold war – between the two great powers of the moment: the Soviet Union and the United States of America.

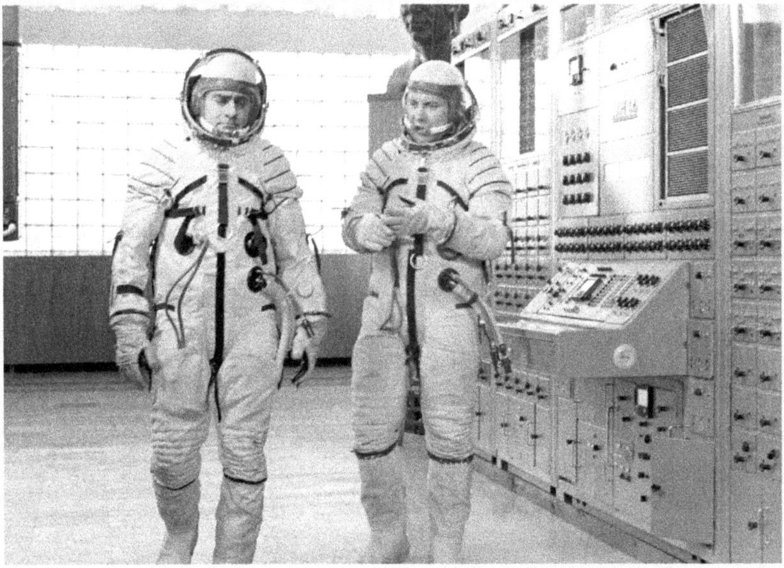

Figure 6.4.2 Ivan listens to a comment from his "shadow"
Source: Joan Fontcuberta

Sputnik: Soyuz 2 Odyssey reassesses the verification of information and the reflection and validation of results. In short, it applies the scientific method to a historical narrative.

The exhibition consists of a collection of photographs and objects that are unique, such as a urine bag used by astronauts in Space, the fork and can opener used on 6 August 1961 by German Titov (the first man to eat in space and the second Soviet cosmonaut), signed photographs, newspapers, technical drawings, diagrams and personal papers belonging to cosmonauts.

Outcomes

The verification of the information, reflection on the truth of the documents or of the missing data, the origin of the information sources, verification of the results . . . all these elements represent the essence of scientific knowledge. In science all information must be verifiable and refutable.

Scientific knowledge progresses because people are capable of doubting established thinking.

The work by Joan Fontcuberta encourages visitors to doubt the information presented.

For the science centre CosmoCaixa this exhibition broadened the range of interested audiences, and the dissemination was extended to people that are normally more interested in "cultural" venues.

Although some museum colleagues did not believe that the exhibition contents were adequate for a science museum, most of the museum staff were ultimately very satisfied. The reaction of the public was positive, except for some visitors who did not think that a science centre should present deception. However, most of the audience felt it was a very positive experience that allowed them to question the truth.

Troubled reactions from staff and visitors speak clearly about expectations that surround science engagement institutions as well as about positivist attitudes. Art allows us to erase the line between science and humanities, broadening the ways we think of each one, challenging and highlighting their internal ontological rules.

6.4.3 Artist scientists

Enfolding artistic and scientific methodology

The scientific method is often described in terms of a sequence of actions: making an observation or identifying a gap in current knowledge; forming a hypothesis or explanation; making a testable prediction; gathering data or conducting analysis to test this prediction; using the obtained results to formulate new predictions. These actions are often complex, iterative, messy, and intrinsically social. Data gathering is often collaborative, organising and analysing data are a contingent and context-dependent process, and the marking of results as 'knowledge' is dependent on the communal processes of recording, disseminating, reviewing, integrating, and replicating.

In certain art-science projects, these "actions" that characterise the scientific method can be brought under the spotlight, allowing direct encounters with the practices of science itself, as well as with some of its "unseeable" objects. Whether an artist deploys scientific tools, a scientist engages directly in artistic practice, or a "hybrid" artist-scientist weaves together these two practices, artistic and scientific methods can be shown in their complexity—and not necessarily contrasting—nature.

Case study: Medical Museion – University of Copenhagen, Copenhagen (Denmark)

Adam Bencard, Associate Professor and Curator
Bente Vinge Pedersen, Head of Exhibitions and Curator
Louise Whiteley, Associate Professor and Curator

Medical Museion is a combined museum and research unit located in the Faculty of Health and Medical Sciences at the University of Copenhagen. Founded on a private initiative in 1907 it now holds one of the richest historical collections of medical artefacts in Europe, and has been working for two decades to bring this history to meet the present and future of biomedicine and health. At Medical Museion the arts have always been seen as one amongst many disciplinary and experiential perspectives on what it is to be a body, healthy or unwell, and what scientific knowledge means for our existential journey through such questions.

1000 Handshakes

A performance experiment by Prof. François-Joseph Lapointe in which he shook hands with 1000 people, assisted by a team of research students who gathered bacterial samples. The samples were analysed to produce 'microbiome selfies'. In this project the framing, the question, the method, and the public outcome were at once both artistic and scientific.

Description

We often think of bacteria as the enemy, but recent estimates suggest that our bodies contain roughly as many microbial cells as human cells.

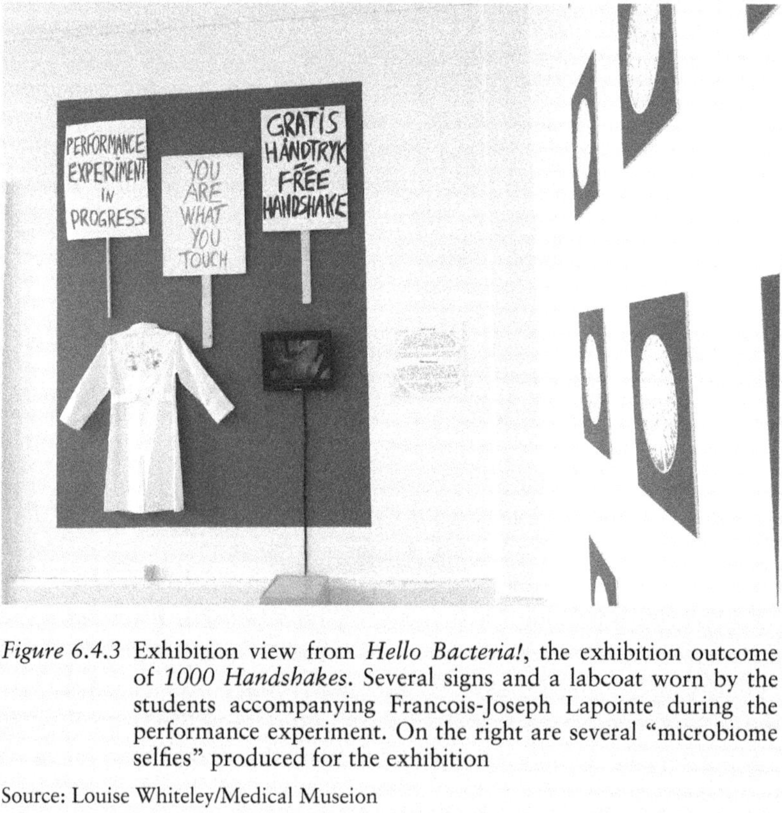

Figure 6.4.3 Exhibition view from *Hello Bacteria!*, the exhibition outcome of *1000 Handshakes*. Several signs and a labcoat worn by the students accompanying Francois-Joseph Lapointe during the performance experiment. On the right are several "microbiome selfies" produced for the exhibition

Source: Louise Whiteley/Medical Museion

Microbes play an essential role in our health and even affect our identity, and every body part has a unique microbial community or 'microbiome'. Every time you shake hands with someone, your microbiomes mingle . . .

In the past decade, the symbiotic entanglement of humans and microbes and the implications for our wellbeing has become a prominent emerging field in the biosciences. Canadian artist-scientist Prof. François-Joseph Lapointe works at the borders of art and science, using his body as a laboratory to explore the transformation of his own microbiome.

In 2014 in Copenhagen, Lapointe collaborated with Medical Museion to produce two versions of the performance experiment *1000 Handshakes*, both times shaking hands with over 1000 people. A team of assistants sampled the microbes on his hand roughly every 50 handshakes, and the data was processed to trace the emergence of an invisible community on his skin. In 2016 the performance led to the exhibition *Hello Bacteria!*, displaying documentation from *1000 Handshakes* and some of Lapointe's 'microbiome selfies', alongside objects from the history of our attempts to control and collaborate with bacteria, and information about current research on the microbiome.

The performance experiment and exhibition were co-developed by the artist and two researchers/curators, Adam Bencard and Louise Whiteley. Eight students from the medicine and arts faculties facilitated the performances and explained scientific ideas to the audience. The collaboration also involved scientists at the Novo Nordisk Foundation Center for Basic Metabolic Research, both as hand-shakers and as partners for sequencing the handshake data.

Outcomes

The performances and exhibition were a great learning experience for the museum in dealing with the complicated field that stands at the border between art, science, performance and curatorial work.

Art collaborations allow topics of shared concern – such as our relationship to microbes on and around us – to be understood, engaged with and discussed in a variety of ways via open-ended exchange across practices and forms of knowledge. This project gave the opportunity to explain some findings that are basic in microbiome research, but in the context of the philosophical and societal questions they raise.

The performances and the exhibition strengthened research collaborations, especially with biomedical researchers who were involved both in the performance and in the attempt to sequence the collected data. It was also a learning experience in the unpredictability of lab work, as the samples from the first performance proved difficult to sequence.

The project encouraged museum staff to continue developing proper entanglements of science, art and cultural history; mirroring the institution's perspective on the relation between these fields 'in the real world'. It reinforced the belief that it is preferable to work in a close and collaborative manner with artists, rather than with more straightforward commissions. Part of the strength of the finished work was that it moved across the three domains, with equal claims of 'ownership' for the project. This allowed the audience to encounter the topic in multiple ways, slipping between the boundaries of science communication, aesthetic experience and reflections on a culturally important health topic.

This project illustrates one specific version of a generalisable approach; allowing arts and sciences to work together to generate the framing or 'question' as well as the methods and outcomes. This working together is sometimes in synchrony, sometimes in conflict or in alternating phases, but resists a clear demarcation of the role of the artist, the scientist, and the curator. Two key practical foundations for such projects are to choose topics that are open-ended in each domain, and to find collaborators who want to bring their discipline into play in producing a *shared* public product; something that won't be judged purely as an artwork, a scientific experiment, or an attempt at scientific dissemination.

6.4.4 Crossbreeding artistic and scientific research

Situations that entail a long-term relationship between artists and scientists – such as lengthy residencies, interdepartmental university courses or interdisciplinary research labs – create the conditions for a cross-pollination of research practices. Scientific research, which is typically shaped around a pre-designed testing of hypotheses, comes into contact with artistic research, which normally works in response to progressive stimuli that shape a nonlinear and unpredictable research path.

For example, in the early 1900s, numerous artists and scientist researched "vision" and crossbred and hacked their reciprocal methodologies. Among the outcomes of these methodological hybridisations in the arts and the sciences were the respective "methodical" products of optical art and the relativistic theories of the Gesthalt, which daringly decided to bring into the scientific experimental practice non-testable aspects of reality such as beliefs and intentions.

Case study: MIT Museum, Cambridge, MA (USA)

Seth Riskin, Manager, MIT Museum Studio and Compton Gallery

The MIT Museum is closely connected with MIT research and education. It includes galleries, offices, collection storage, and education spaces as well as two on-campus spaces. It holds just over a million objects in its collections and offers talks, workshops, family events, and other programs.

Vision in Art and Neuroscience

An MIT course offering an experimental exhibition setting for an in-depth exchange between art practice and scientific theory of vision. The course involved artists, scientists and museum professionals in the articulation of a model of mutually informed theory and practice of vision that was presented in a public exhibition, which also triggered reflections on current museum practice.

Description

Funded by the MIT Center for Art, Science & Technology with a Mellon Faculty Grant, the course was offered to undergraduate and graduate MIT students from all disciplines through the MIT Department of Brain and Cognitive Sciences. The course took place during fall semester 2017 and was a collaborative effort between Pawan Sinha, MIT Professor of Vision and Computational Neuroscience, Sarah Schwettmann, PhD student of MIT Department of Brain and Cognitive Sciences, and Seth Riskin, artist and MIT Museum Studio Manager.

Vision in Art and Neuroscience was inspired by the ground-breaking book and exhibition *Illusion in Nature and Art* by Richard Gregory and Ernst Gombrich, that presented and probed the reciprocal relationship between optical information in the eye and knowledge of the world in the mind. Contemporary inspiration came from the work of artist James Turrell, who crafts light and the material environment so as to remove the

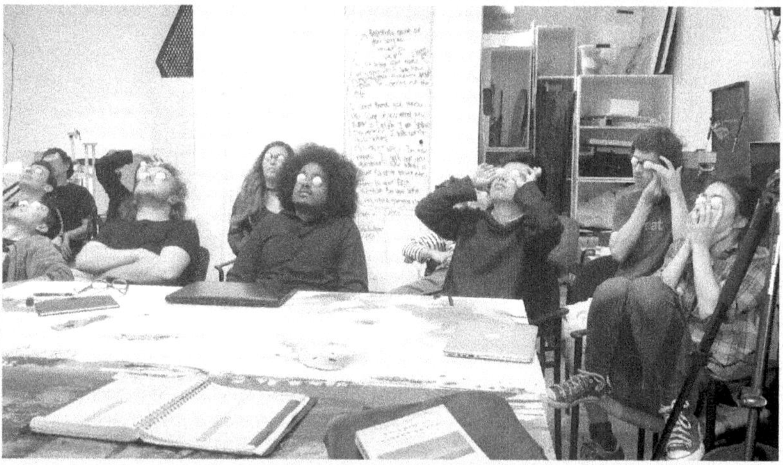

Figure 6.4.4 Students explore the perceptual experience of a ganzfeld, an equiluminant field, via a simple light diffusion mechanism created in class: ping-pong balls cut in half

Source: Photo by Sarah Schwettmann

scaffolding of perceptual habit and activate awareness of the rudimentary and creative dimensions of the seeing process.

The course explored the intersection between art and neuroscience in a novel manner: by treating perception as an act of creation (the creation of an individual's world of experience) and finding shared ground between art and neuroscience in the creative faculty of vision. Students learned to develop and use new tools in the arts (working mostly with light) and vision neuroscience. In the MIT Museum Studio and Compton Gallery, 14 students, from areas such as brain and cognitive sciences, mathematics, mechanical engineering, and architecture, developed individual artworks. Composing an exhibition titled *Perceiving Perception*, the artworks allowed members of the MIT community and the visiting public to perceptually experience the levels of hierarchy of visual processing discussed in the course.

Outcomes

It is increasingly clear to vision science that some visual artists execute a subjective study of vision in their work, probing the interplay between the visual brain and the physical world and manifesting internal visual processes. This presents an opportunity for collaboration: for neuroscientists, visual art can serve as a window onto visual processing and, in turn, neuroscience can guide and propel artistic exploration and production.

For the MIT community, the exhibition *Perceiving Perception* provided a model of integrated art and science. The projects on display revealed areas of future artistic exploration, accessible with knowledge of early visual processing, equipment and techniques for shaping visual perception. Equally, the experiential knowledge embodied in the artworks showed potential of contributing to vision theory, posing new questions and possible advances.

The primary motivation of the course was thus to achieve a new level of interaction between museum practice, art and science through the subject of human vision. Art provides method and works for *Perceiving perception*, applicable to the advance of both art and science.

Visual intelligence, just like most fields of reality, can be studied through combined disciplinary lenses that overstep academic disciplinary divisions. Trans-disciplinary experiences offer researchers from all backgrounds the opportunity to focus on their own ground of interest while applying methodologies and stimuli from other fields of knowledge.

6.4.5 Questioning common museological practices

Common museum practices themselves can be questioned. How and what do we collect? What are the underlying hierarchies of objects and concepts in the physical layout of an exhibition? Can an object make us laugh, cry, enrage and can we understand why it does this?

The making, the history, the preservation, and the interpretation of objects is the realm of material culture studies, yet the reflection can be also brought to life through the processes and products of artistic research.

Once again contemporary artists can play a revealing role, uncovering unconscious or unchallenged museological practices such as the way we display certain objects that pertain to spiritual spheres or the provenance of our collections.

Case study: Museum für Naturkunde Berlin, Berlin (Germany)

Anita Hermannstädter, Head of Department History of Science & Museum Studies

The Museum für Naturkunde Berlin is an integrated research museum within the Leibniz Association. With over 30 million items, it is one of the most important research institutions worldwide in the areas of biological and geological evolution and biodiversity.

Art/Nature

A four year international pilot research project to create an experimental space for interaction between art, museum praxis, and research into nature. The resulting exhibitions and art projects challenged some of the Museum's own curatorial and exhibition practices, suggesting new ways to think of topics in permanent galleries.

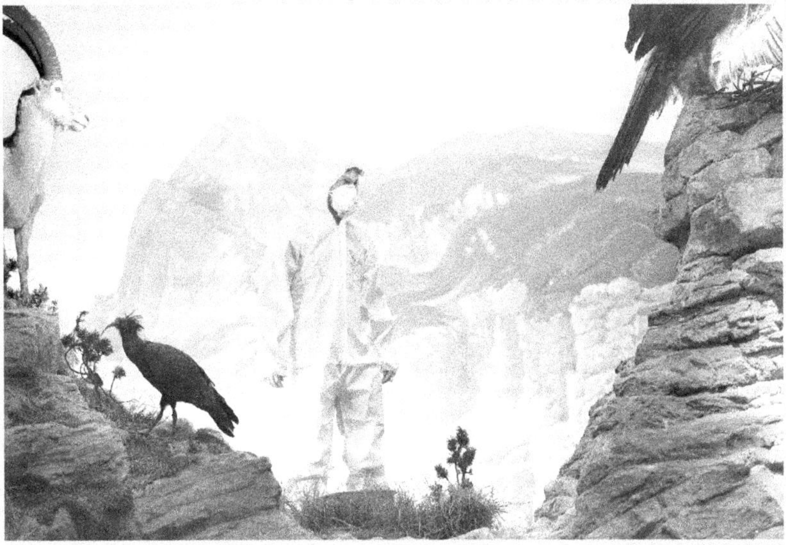

Figure 6.4.5 Video artist Assaf Gruber in a historical diorama of the Museum für Naturkunde Berlin at the screening for his film The Conspicuous Parts

Source: Photo by Hwa Ja-Götz, Museum für Naturkunde Berlin

Description

What happens when contemporary art encounters a natural history museum? This was the experimental spirit that shaped the project in its research on interactions between contemporary artists and a natural museum setting. External curators created curatorial concepts and selected fitting artists. Then the writers and visual/sound artists were given *carte blanche* to explore and interpret the natural history museum by creating artworks. The artists met staff from different museum departments to learn about the museum and present their own ideas. Collaborations ranged from a research trip to record fish sounds to letting the mounted specimens inspire poems about other species' senses. Some staff participated first-hand in performances or provided material for the artworks, whilst other organised and held public Art/Nature talks with museum scientists, curators and artists during exhibition periods.

The works – a sound poem for the bird collection, a lost-and-found for the evaluation of findings, a photo booth that shows how animals see us, etc. – were displayed in turn in the museum galleries, so that regular museum visitors could access at no extra cost the four consecutive

Art/Nature exhibitions with works by: A K Dolven, Saâdane Afif, Sabine Scho and Andreas Töpfer; Serotonin, Fernando Bryce; Klara Hobza, Monika Rinck; Mark Dion, Assaf Gruber, Ulrike Haage and Mark Ravenhill, Elizabeth Price. The project was funded by the German Cultural Foundation.

Outcomes

The aim of the project was to develop new perspectives on the Natural History Museum by questioning its collections, its research and exhibiting practices through artistic methods and expressions. The idea was that this project would provide a contribution to the reflection on the cultural and political role of natural history museums in general and to the discussion about relationships between science and society.

Contemporary art deals with current social and political themes, it offers another aesthetic frame and proposes experiences, that can add to, or contrast, science communication in a natural history museum.

The artistic interventions transformed the museum into a stage for the intersection of science and art, they revealed hidden (his)stories and addressed the multi-layered nature of the museum.

Evaluated by an external organisation, the response of the general audience to the introduction of contemporary art seemed very positive and increasing over the course of the project. The museum managed to reach out to those interested in art and culture, which was a major contribution to audience development.

Art interventions can open up political and poetic dimensions in playful, ironic and intellectual ways. New or often overseen topics and odd materials may take the centre stage of a museum and new aesthetic experiences, such as sound installations or performances, can change the perception of the museum space itself. Above all, the specific character of contemporary art – ambiguous, rich in associations, even enigmatic – can stimulate imagination and trigger deeper reflections and dialogue.

6.4.6 Tackling "cold" topics through emotions

When we are faced with a work of art we are also faced with a set of reactions that can range from liking to disliking. Yet we can also feel other types of emotions such as rage, fear, confusion, irritation, happiness, euphoria. Among these emotions there are some that tend to stimulate further thinking and learning: interest, confusion and surprise. Interest occurs when the artwork or item we see fits into our knowledge and expectations while providing a new experience. Confusion occurs when we cannot understand what we are witnessing so that we are somehow compelled to solve this situation by seeking complementary information. Surprise occurs when something unexpected disrupts our actions or senses and we are forced to turn our attention to an incongruous item that we will then tend to investigate.

If we stir these emotions around a topic that usually leaves us more detached, we can create an interpretative inversion and spark interest at a more personal level.

Case study:
Quai des Savoirs, Toulouse (France)

Marina Leonard, Program Manager

Quai des Savoirs is a platform devoted to sharing scientific, digital, technological and industrial culture in Toulouse. It provides a multi-disciplinary programme with exhibitions, workshops, "coffee-house" debates, conferences, mediation and shows. It opened its doors in Toulouse in 2016.

Le Synesthesium

An immersive sound experience, an artistic performance and a set of workshops that approach an area of specialisation of the city of Toulouse:

aeronautical research and industry. The project offered new ways of approaching a technical theme and it was the perfect excuse to create new bonds with research and industry partners.

Description

Singer and writer Arthur H and his partner sound designer Léonore Mercier created a unique performance under a large geodesic dome. The artists were identified among the participants in a one-month thematic scientific and cultural festival related to flight, which took place in 2016 and included exhibitions, workshops, meetings with artists and scientists and art and science projects (installations, live entertainments, workshops).

To enrich the Synesthesium experience – programmed purposely to close the festival – the artists and Quai des Savoirs decided to collaborate and create a special play based on sounds captured in Toulouse. So Léonore went "sound hunting" for several weeks, aided by the centre's staff in shaping the play and accessing restricted areas of the airport as well as of other industrial, scientific, cultural and educational settings.

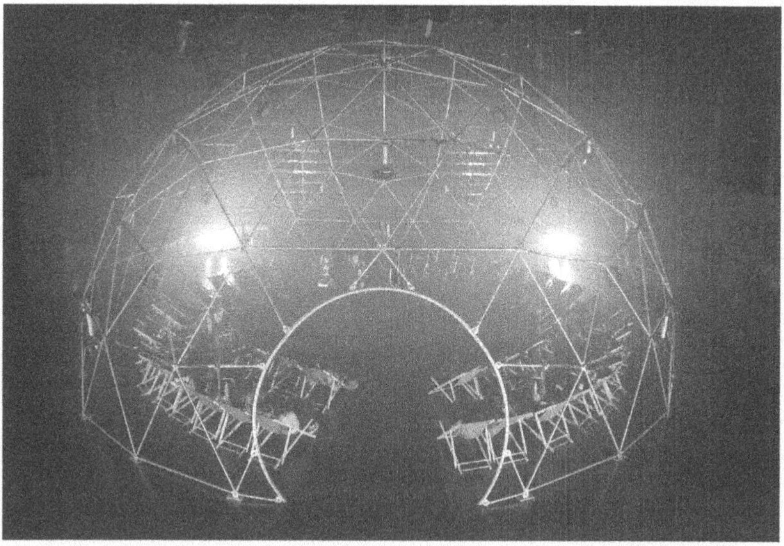

Figure 6.4.6 Synesthesium, the geodesic dome
Source: Photo by Mathilde Delahaye

The result was a programme that included different experiences for visitors:

- a day-long opening of the dome to experience the 15 minute *Toulouse play* as a poetical immersion in one of the main areas of excellence of Toulouse: aeronautic industry. This immersive setting encompassed a recorded sound composition broadcasted by 24 loudspeakers
- live performances of the artists under the geodesic dome, echoing the festival's cultural and scientific program twice a day. This setting encompassed a live sound composition by Léonore while Arthur H recited his poems dedicated to travelling, airports and flight;
- one special program dedicated to the partners of the creation and very close partners of le Quai des Savoirs to discover the 15 minute *Toulouse play* and the live performance with the artists;
- four workshops for young people where Léonore guided participants in the creation and composition of a play using the unique sound system of the dome.

Outcomes

Programs on aeronautical research and industry – such as exhibitions and conferences – are very frequent in Toulouse, but most of the time they are very technical. The idea behind this collaboration was to present the subject in a totally different way.

Thanks to an original technical device such as the geodesic dome, with its unique sound sculpting system, visitors were presented with an alternative way of thinking, another point of view, the possibility to discover a scientific concept in a poetical way. Students participating in the workshops felt empowered by the fact that after hosting Léonore in their world while she was collecting sounds in their school, they could then access hers, in a mutual exchange of scenarios and competences.

For the personnel of the Quai it was a great opportunity to discover and participate in a high level artistic creation, accepting a change of perspective on the way to present a topic. In fact, it was interesting to work with the artists on the atmosphere of the narrative and on the industrial and educational locations that could be involved. In the end, the selected research, cultural and industry partners that agreed to welcome Léonore grew more confident towards the quality and scope of the project which ultimately lead to the creation of new institutional relationships.

Working with artists on relevant or controversial topics gives the opportunity to add a poetical dimension to a reality that can be normally perceived as "cold" and distant. This emotional involvement is a precious basis for deepening understanding and developing dialogue and critical thinking.

6.4.7 Beyond answers

Art and good questions

Hybrid institutions that stand at the crossroads between art and science engagement can become interesting spaces for debate. Here artistic provocations are balanced by scientific explanations or ongoing research.

The silent contract with audiences of such places is that they can provide a new platform in which futuristic or out-of-the-box thinking is balanced by facts.

The starting point are not necessarily science and technology but rather mankind with its many facets. The challenge is to look at these topics through multi-layered interpretations that involve both imagination and rationality. The showing of a combination of artistic and scientific work reveals an attempt not only to offer good answers – as many science centres and museums do – but also to provide good questions – which is something that is exceptionally accomplished by much contemporary art.

Case study:
Science Gallery Dublin, Dublin (Ireland)

Ian Brunswick, Head of Programming

An initiative of Trinity College Dublin, Science Gallery Dublin opened in 2008. It organises temporary, surprising and thought-provoking exhibitions and events fuelled by the expertise of scientists, researchers, students, artists, designers, inventors, creative thinkers and entrepreneurs.

Grow your own

An exhibition created by Science Gallery at Trinity College Dublin inviting visitors to consider some of the potentially ground-breaking applications and uncertain implications of synthetic life. The exhibition brings together actual artefacts and speculative design, challenging viewers to relate and make sense of these two areas of reality in the construction of their views of science as a product of and for society.

Description

Science Gallery's 2013 exhibition *Grow Your Own* dealt with the emerging field of Synthetic Biology by exhibiting artworks and designs that portrayed 'speculative futures' on how we might utilise genetic modification. GYO was curated by a mix of synthetic biologists, designers, researchers and artists, and the exhibition was staffed by Science Gallery 'Mediators' (primarily Trinity College Dublin students).

Displays of the exhibits were deliberately ambiguous, inviting visitors to investigate and evaluate how realistic each piece was. Some found it confusing, but the theatricality and absence of overt signage resulted in a 'suspension of disbelief'.

This is best exemplified by the work *I Wanna Deliver a Dolphin* by Ai Hasegawa which described a future mechanism for humans to give birth via an engineered hybrid placenta to threatened mammalian species.

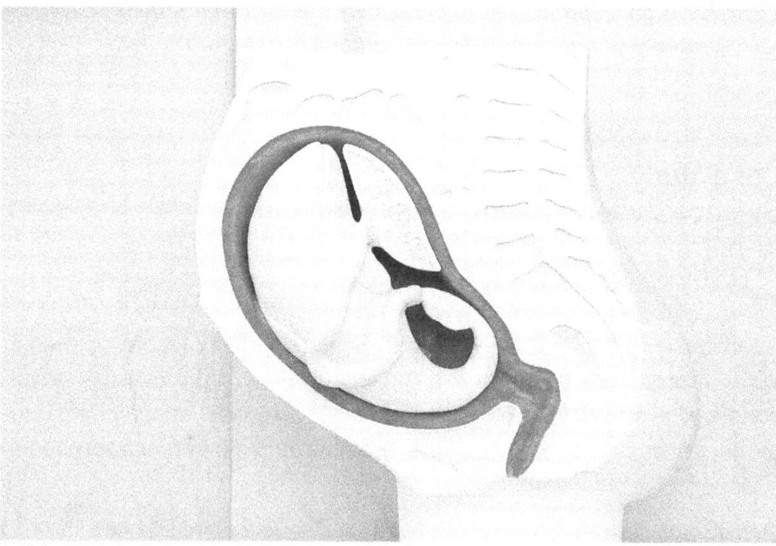

Figure 6.4.7 I Wanna Deliver a Dolphin by Ai Hasegawa at GROW YOUR OWN at Science Gallery at Trinity College Dublin, 2013

Source: Science Gallery Dublin

The hyper-real installation included a model of a modified human placenta, description of the necessary genetic modifications, and a video of the simulated interspecies birth.

Mediators were on hand to discuss how visitors felt about a future where humans could birth endangered species – something hard to imagine without such a visceral portrayal.

Outcomes

The goal of the exhibition was to anticipate, imagine and debate applications of synthetic biology before they became scientifically feasible. Exhibits ranged vastly in realism: some were actual artefacts, while some were hypothetical designs which awaited necessary (but likely) technological advances. Others were imaginary, unrealistic artworks, portraying an alternate reality to mirror and illuminate our own.

In the absence of mediators (such as on the YouTube video about the work) some viewers could not properly distinguish between facts and fiction. Some believed I Wanna Deliver a Dolphin was real, and voiced vehement objection.

Instrumentalisation of art for the agenda of science is problematic, but one important value of art is that it often beholds an emotive quality.

Furthermore, artists possess a different social contract with the public than scientists. Ideally, you should never have to ask, 'is this real?' about science, yet it can be a powerful tactic of art to fool you into thinking something is real, or to make something feel real even when you know it is not. This is why we don't intervene to stop a murder during a performance of Shakespeare. Yet it emotionally affects us. We suspend our disbelief.

Deception of this kind is both powerful and hazardous. The artworks in GYO had profound effects on people, evoking strong moral and ethical consideration precisely because for a moment the visitors suspended their belief. The artworks imagined a future, and the visitors played along, discovering their true feelings about a fictional world which science may soon make possible.

When confronted with artworks within "science" exhibitions, people tend to engage differently with displayed items. For example, the fact of not being sure whether something is real or not can help visitors discover their 'gut reaction' to particular technological applications. This is a powerful way to develop deeper thinking, triggered by personal ethical positions that might be unveiled by a peculiar situation that mixes facts and imagination in unexpected and non-immediately identifiable ways.

6.4.8 Art connecting personal meanings and collective standpoints

Many contemporary artists produce works that become epicentres of debate, tackling ethical issues that relate the ideas of individuals to those more broadly accepted by society.

So-called community-based art, interactive art and participatory art are some of the artistic strands that have related personal views and reactions of the audience to broader social topics.

This connection between collective narratives and personal perspectives tends to transform the subject of the artwork into something deeply relevant for the individual viewer, affecting his/her view of society.

Case study: Wellcome Collection, London (UK)

Kate Forde, Head of Exhibitions
James Peto, Head of Programme

Wellcome Collection is a free museum and library that aims to challenge how we all think and feel about health. It creates opportunities for people to think deeply about the connections between science, medicine, life and art. It is part of Wellcome, a global charitable foundation, which aims to improve health for everyone by helping great ideas to thrive.

Serena Korda: Laid to Rest

A commissioned installation which formed part of the exhibition *Dirt: the filthy reality of everyday life.* The installation was accompanied by a series of performances, a film, a procession, a vinyl record and a permanent public memorial to the project. The project was managed and produced by UP Projects. It represented a vital contribution to the exhibition's discussion on Dirt in a local, historical context, but brought the narrative firmly into the present and into the daily lives of local people.

Description

Laid to Rest was commissioned as part of the 2011 exhibition *Dirt: the filthy reality of everyday life.* Dirt poses a significant risk to our health but is also vital to our existence. Following anthropologist Mary Douglas' observation that dirt is "matter out of place", the exhibition looked at scientific and cultural attitudes towards dirt in six very different places: from 17th-century Holland to a New York landfill site in 2030.

Wellcome Collection is situated close to Euston's "brick fields" and what was, in the first half of the 19th century, the "Great Dust Heap" of London. Ash from the dust heap was mixed with Euston clay for baking into London bricks. The artist Serena Korda collected dust from 450 members of the public and baked each sample into a brick, marked with the donor's initials. With the initials facing outwards and accompanied by a list of the donors, the block of bricks had an imposing presence in the exhibition.

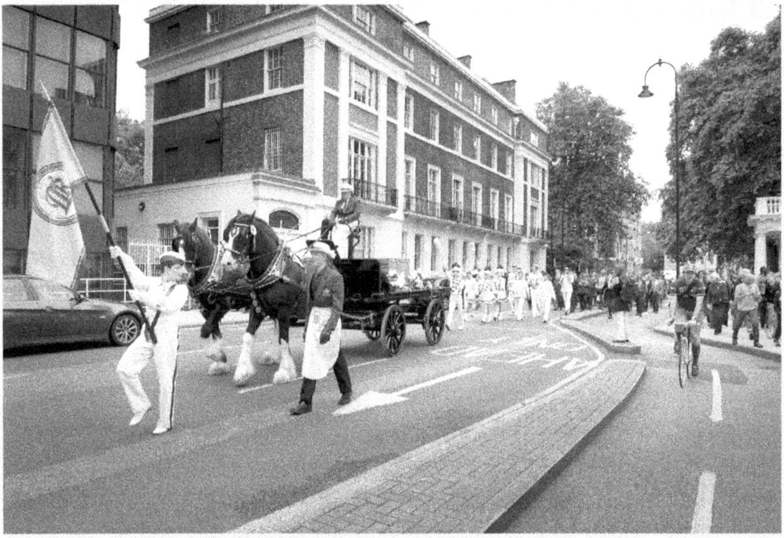

Figure 6.4.8 Laid to Rest, 2010–2012 by Serena Korda, commissioned by the Wellcome Trust in association with UP Projects as part of the Wellcome Trust's DIRT season

Source: Photo by Dave Sayer

A series of choreographed performances accompanied Laid to Rest – one filmed at a brick-built sewage pumping station. After the exhibition a horse-drawn procession took the bricks to be re-buried in London soil. A plaque commemorates the burial site in Brunswick Square Gardens.

Outcomes

Because the exhibition was in six sections, each focussing on a different geographical location and a different time in history, we wanted to include a significant new work which would be specific to our own time, our location on Euston Rd. and our immediate audience. We wanted the commission to explore the continuing complexities of our relationship with dirt, which we spend so much time trying to get rid of, but which is also essential to our lives. Context is all.

Laid to Rest made a vital contribution to the exhibition's emphasis on the importance of context to scientific and cultural understandings of dirt. Serena Korda's work was part of the section that looked at Victorian London, where the task of discovering the cause of cholera was an urgent one. Taking the historical significance of Wellcome Collection's location as her starting point, she looked at Euston's role in the story of what London did with its dirt, but she brought that story firmly into the present day, involving many of our visitors and neighbours who contributed dust from their own properties.

In an exhibition that looked at the relationship between health and sanitation in the past (Victorian London), present (urban India) and future (New York City), Korda's work wove together ideas about 'soil' and 'dust', about cleaning, recycling, and burial, encouraging fresh ways of thinking about what dirt is, where it comes from, and its possible uses. This is a subject for artists, sociologists and anthropologists as well as for scientists, health workers and urban planners.

Artists' contributions to exhibitions in institutions rooted in science can greatly extend visitors' engagement with the subjects the exhibitions address. They can do this by encouraging active participation, by extending the experience beyond the exhibition space, or by helping visitors find personal connections with context and subjects they might otherwise have thought of as transcending their field of interest or being far from their everyday life.

7 (open) Conclusions

The post-disciplinary museum

Camilla Rossi-Linnemann and Giulia de Martini

It has become clear that much scientific museology has moved from earlier notions that art should primarily serve as a medium to illustrate/promote science to the inclusion of art as a more integral element of its practice.

Let us retrace, briefly, some of the main concepts emerging from the experience of the authors contributing to this volume.

First of all, there appear to be two more "conventional" ways to incorporate art, which do not usually challenge traditional science curation. In the first case, science curators, educators and designers "use" art for its aesthetical qualities, to provide science exhibitions or concepts with intriguing "forms". In the second case, art is used as a sort of "translation" tool, that aids to "visualise" through metaphors certain aspects of the world that would otherwise be ungraspable or inexpressible with the traditional narrative media of a science exhibition/experience.

In other instances, artists are asked to contribute in the design of interactive exhibits, or in the planning of activities, or by directly involving visitors in the creation of artworks and performances. In this case artists are seen more as "enablers" than as "translators". It is in fact commonly recognised that art can play an active role in the learning process, triggering emotional response, stimulating the connection of contents to individual beliefs, kindling open-ended creativity and imagination, and ultimately sustaining individuals in valuing active thinking.

Last but not least, the peculiar ability that artists have to see the world differently, to imagine future scenarios, to unveil concealed and often unconscious biases or ethical frames, has often placed art in the unique position of encouraging the debate and reflection on science and technology itself and on the ethical issues raised by new products of scientific and technological research.

Yet the emotional power of art must be considered not only for its positive but also for its more challenging aspects, which is the reason why many science museums still make a "safe" use of artists as external suppliers rather than involving them more deeply in their everyday practice. In a world where emotions seem to rule over facts, numerous science engagement institutions have taken it upon themselves to "protect" the image of neutrality, objectivity and universality of science. Bringing art into the equation is seen by many as a way to dangerously relativise science. Yet science is, in fact, a cultural product, and many of the institutions presented in this book have chosen to bring this into the open through art projects that strengthen the scientific discourse precisely by placing in the spotlight its values, weaknesses, and connections to ethics and other aspects of human knowledge.

Just as in any successful relationship it is important to build elements of reciprocal trust and find the right partners. Authors underline the importance of working with artists who are interested in the research process and not only in its content, media and output. They also recommend involving artists (and science communicators) who take into consideration audiences as an integral and active element of their practice, trusting them with the ability to participate in the making of new cultural products.

The visionary idea of a "total museum" described by Jorge Wagensberg has inspired contemporary museography in many ways, whilst the notion of "museum" has increasingly called for new and more active forms of engagement. Museums have always presented themselves as places offering access to knowledge, yet knowledge, as Mark Wright from FACT points out in his chapter in this volume, is not an artefact, it is "performative", it must be experienced and played with in order to exist. And somehow artists seem to have the unique ability to build creative environments and conditions where the inputs of scientists, philosophers, anthropologists, and many others can be more easily taken, hacked, remodelled and used by others (i.e. visitors) in their own lives.

Moreover, as mentioned here by Maria Pfeifer and Christopher Lindinger from Ars Electronica, artists often work with an imaginary dimension that helps 'escape predeterministic views of science and technology', ultimately showing that it is the way in which each one of us creatively thinks about science and technology (and about its connections with ethics, economics and many other aspects) that opens up concrete social scenarios that depend on individual choices and engagement.

Adopting Latour's wonderful description of his own exhibitions,[1] we highlight the value of curatorial experiments that encompass heterogeneous contributions, so that rather than as a outputs of precise purposes, the results of these collaborations can be described as interference patterns – which can be both constructive and destructive. In this sense, the exhibitions themselves become rich cultural products that include in their making all different factors involved – from curators to audiences, from captions to objects, from interpretation to buildings, to all other implicated media.

Today museums are striving to incorporate a plethora of different views. We believe that wherever this process is left to mature and is not hindered by rigid traditional views, museums will spontaneously shift from being centres for the communication and understanding of separate disciplines to *post-disciplinary centres* that aim to empower audiences and foster the blended making of new knowledge and cultural products.

Overall, there is no question about the fact that among institutions that recognise themselves as actors in the field of science engagement there is a growing impulse to turn to art in a variety of different ways. These contemporary museums are reflecting the spirit of our time and the way disciplines are combining in the outside world. The term *post-disciplinary museum* ultimately describes institutions that actively explore new ways to communicate, engage and produce knowledge.

Note

1 Weibel, P. & Latour, B. (2007) Experimenting with Representation: Iconoclash and Making Things Public, in: Macdonald, S. & Basu, P. (Eds.) *Exhibition Experiments*, 94–108, Hoboken, NJ: Blackwell Publishing Ltd.

Index

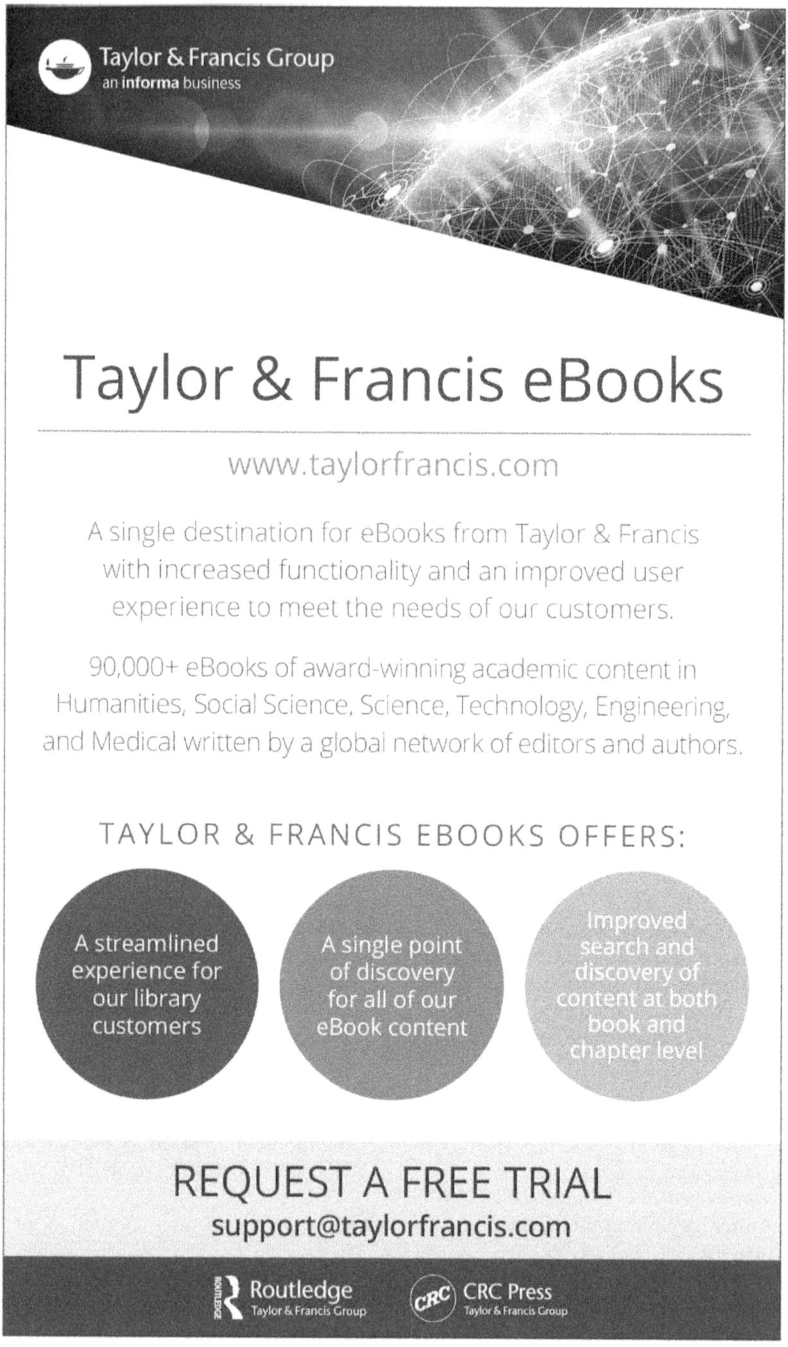
Taylor & Francis Group
an informa business
Taylor & Francis eBooks
www.taylorfrancis.com
A single destination for eBooks from Taylor & Francis with increased functionality and an improved user experience to meet the needs of our customers.
90,000+ eBooks of award-winning academic content in Humanities, Social Science, Science, Technology, Engineering, and Medical written by a global network of editors and authors.
TAYLOR & FRANCIS EBOOKS OFFERS:
A streamlined experience for our library customers
A single point of discovery for all of our eBook content
Improved search and discovery of content at both book and chapter level
REQUEST A FREE TRIAL
support@taylorfrancis.com
Routledge
Taylor & Francis Group
CRC Press
Taylor & Francis Group

For Product Safety Concerns and Information please contact our EU
representative GPSR@taylorandfrancis.com
Taylor & Francis Verlag GmbH, Kaufingerstraße 24, 80331 München, Germany

www.ingramcontent.com/pod-product-compliance
Ingram Content Group UK Ltd.
Pitfield, Milton Keynes, MK11 3LW, UK
UKHW010956180926
475701UK00004B/396
9781032082974